Shark Repellents from the Sea

New Perspectives

AAAS Selected Symposia Series

Shark Repellents from the Sea

New Perspectives

Edited by Bernard J. Zahuranec

Routledge
Taylor & Francis Group
LONDON AND NEW YORK

First published 1983 by Westview Press

Published 2019 by Routledge
52 Vanderbilt Avenue, New York, NY 10017
2 Park Square, Milton Park, Abingdon, Oxon OX14 4RN

Routledge is an imprint of the Taylor & Francis Group, an informa business

Library of Congress Catalog Card Number 83-60529

ISBN 13: 978-0-367-28720-7 (hbk)

About the Book

Some species of sharks, because they are large, agile predators, are extremely dangerous to humans in water. During World War II, the long-time goal of an effective shark repellent was partially realized by the U.S. Navy's development of "Shark Chaser," a copper acetate-negrosine dye mixture. But neither Shark Chaser nor other noxious or even extremely toxic chemicals have provided fully dependable, long-lasting repellent action in the open ocean. Recently, however, largely as a result of research on the glandular secretions of a Red Sea flatfish--the Moses Sole, *Pardachirus marmoratus*, there is new hope of developing a truly effective shark repellent, using the principles embodied in naturally occurring, biologically active substances.

This book deals with the biochemistry and physiopharmacology of these naturally occurring substances. The contributors examine the effects of the substances on the behavior of sharks and show how solutions to the problem of repelling sharks can pave the way to a much broader understanding and wider uses of these and similar substances.

About the Series

The *AAAS Selected Symposia Series* was begun in 1977 to provide a means for more permanently recording and more widely disseminating some of the valuable material which is discussed at the AAAS Annual National Meetings. The volumes in this *Series* are based on symposia held at the Meetings which address topics of current and continuing significance, both within and among the sciences, and in the areas in which science and technology impact on public policy. The *Series* format is designed to provide for rapid dissemination of information, so the papers are not typeset but are reproduced directly from the camera-copy submitted by the authors. The papers are organized and edited by the symposium arrangers who then become the editors of the various volumes. Most papers published in this *Series* are original contributions which have not been previously published, although in some cases additional papers from other sources have been added by an editor to provide a more comprehensive view of a particular topic. Symposia may be reports of new research or reviews of established work, particularly work of an interdisciplinary nature, since the AAAS Annual Meetings typically embrace the full range of the sciences and their societal implications.

WILLIAM D. CAREY
Executive Officer
American Association for
the Advancement of Science

Contents

About the Editor and Authors

Bernard J. Zahuranec, *a specialist in fish systematics and zoogeography, is program manager of the Oceanic Biology Program, Office of Naval Research Detachment, NSTL, Bay St. Louis, Mississippi. He has been managing ONR basic research contracts on dangerous and noxious marine organisms, including sharks, since 1975, and he is the editor of "Shark Research: Present Status and Future Direction" (ONR Report ACR 208).*

Gerald J. Bakus, *a marine ecologist, is associate professor of biological sciences at the University of Southern California. A specialist in predator-prey interactions, marine toxins, and chemical defense mechanisms, he also serves as executive officer of the International Environment Organization in India.*

H. David Baldridge, Jr., *currently senior research associate at Mote Marine Laboratory in Sarasota, Florida, has been involved in research on shark deterrents and hazards of shark attack since 1964. Formerly coordinator of shark repellent testing for the Office of Naval Research, the American Institute of Biological Sciences, and the Shark Research Panel, he is the author of numerous scientific and popular publications on sharks, shark attack, and antishark measures, including* Shark Attack *(Droke House/Hallux, Inc., 1974).*

Yechezkel Barenholz *is professor of biochemistry at Hebrew University Hadassah Medical School in Jerusalem, Israel. A specialist in lipid enzymology and membrane biophysics and biochemistry, he is the author of more than ninety publications in his field.*

Celia Bonaventura, *codirector of the Duke University Marine Biomedical Center in Beaufort, North Carolina, is* *interested in protein engineering and technology. Her research and publications have focused on hemoglobins, hemocyanins, cytochrome oxidase, and the structure-function relationships in respiratory proteins.*

Joseph Bonaventura *is codirector of the Duke University Marine Biomedical Center in Beaufort, North Carolina, where he is concerned with the use of marine species in studies involving the health of humans and their environment. His published work includes studies on hemoglobins, cytochrome oxidase, and the structure-function relationship in respiratory proteins.*

Eugenie Clark, *professor of zoology at the University of Maryland at College Park, has been actively studying the behavior and taxonomy of marine fishes in general and the behavior of sharks in particular for many years. Currently she is also a research associate in ichthyology at the American Museum of Natural History in New York City. She has published numerous articles and books, both scholarly and popular, and has received many honors and awards for her work.*

Samuel H. Gruber *is associate professor of marine science at the Rosenstiel School of Marine and Atmospheric Sciences, University of Miami in Miami, Florida. A specialist in sensory physiology, behavior and physiological ecology of marine organisms, he has been doing laboratory and field studies of sharks involving whole organism energetics, population dynamics, and behavior.*

Joseph V. Martin, *a neurochemist, is a staff fellow at St. Elizabeth's Hospital, National Institute of Mental Health, Washington, D.C. His research has focused on the effect of extracts of marine organisms and of liposomes on neurotransmitter release.*

William O. McClure, *a neurochemist, is professor and director of the Neuroscience Program at the University of Southern California. His research interests include study of neurotoxic activities from marine organisms and regulation of cholinergic function in the brain. Currently he serves on the editorial boards of several journals, among them the* Journal of Neurochemistry *and the* Journal of Neuroscience Research.

Donald R. Nelson, *professor of biology at California State University at Long Beach, has been studying the behavior and sensory physiology of sharks for many years. He is well known for his work on acoustic attraction, hearing thresholds, agonistic display and attacks, movement patterns, schooling behavior, interactions between divers and sharks, and the development of telemetering techniques for studying free-ranging sharks. In 1980 he was given the James Dugan Award by the American Littoral Society for outstanding service to marine science.*

Naftali Primor *is a marine toxicologist at Osborn Laboratories of Marine Sciences, New York Aquarium, Brooklyn, New York. A specialist in animal toxins and ion osmoregulation in fish, he has done extensive research on the effects of pardaxin on sharks, including its conformation, toxicity, and cytotoxic action.*

Anthony T. Tu, *professor of biochemistry at Colorado State University at Fort Collins, has been studying the conformation of venoms in general and pardaxin in particular. He is the author of* Venoms: Chemistry and Molecular Biology *and* Survey of Contemporary Toxicology, Vols. 1 and 2 *(all published by Wiley, 1977, 1980, and 1982, respectively),* Rattlesnake Venoms: Their Actions and Treatment *(Marcel Dekker, 1982), and* Raman Spectroscopy in Biology: Principles and Applications *(Wiley, 1982).*

Jose A. Zadunaisky *is director of Sackler Institute of Graduate Biomedical Sciences, professor of physiology and biophysics, and professor of experimental ophthalmology at New York University Medical Center in New York City. He is the author or editor of numerous publications in his field, most recently* Chloride Transport in Biological Membranes *(Academic, 1982), and he is coeditor of three volumes in the* Current Topics in Eye Research Series *(with H. Davson; published by Academic Press in 1979 and 1980).*

Eliahu Zlotkin, *associate professor of zoology at Hebrew University in Jerusalem, Israel, has been doing research on the toxic and antitoxic factors in the skin secretion of* Pardachirus marmoratus. *A specialist in insect physiology and toxicology, he has discovered a new group of toxic proteins derived from scorpion venoms which selectively affect insects.*

Preface

Sharks, shark research, and all aspects of shark biology hold great fascination for us, although despite a great deal of speculation, no one knows why. It cannot be merely that sharks kill humans because the number of deaths in the United States attributable to sharkbite is miniscule, probably ranking below lightning and bee-stings as a cause of death. Nevertheless, our fascination with sharks extends even to seeking a means of protection from them.

At least one chemical shark repellent--Shark Chaser, which was developed during World War II (see chapters 1 and 2)--has been widely used in the United States. This compound, however, as well as all other chemicals and repellents tested, is not completely effective. We may now be on the threshold of finding true shark repellents out of Nature's cornucopia of biologically active substances, that catchall phrase for chemicals produced by organisms that affect the physiology or behavior of other organisms.

With this potential in mind, a group of interested scientists met in Toronto, Canada, on 5 January 1981 to take part in a symposium during the annual meeting of the American Association for the Advancement of Science (AAAS). A number of biologists of varied backgrounds, interests, and training attended, which provided an opportunity for an interdisciplinary exchange of information and viewpoints both stimulating and beneficial to all. The contributions to this volume are essentially the papers that were delivered at the symposium, subsequently revised and expanded to include more recent information.

One serendipitous, but not altogether unexpected, outcome of the symposium was the realization that, though it may be intrinsically useful to develop a truly effective shark repellent from naturally occurring, biologically active substances, there could ultimately be greater benefits. For example, once we understand how a biologically active compound works, we might discover other, more important, uses for it. Alternatively, the entire search for, and development of, a shark repellent could be a useful paradigm for finding other kinds of useful biologically active compounds.

Of course, there were many people instrumental in the success of the symposium. Walter Chavin of Wayne State University, Secretary of the AAAS Biological Sciences Section, first suggested the symposium series as a possible mechanism for holding such a meeting and provided a number of helpful suggestions. Mary-Frances Thompson of the American Institute of Biological Sciences helped with travel arrangements for the scientists who attended. John and Sarah Stoneman of Toronto made some of the local arrangements; in particular, Sarah Stoneman obtained additional copies of material that participants had not received in time for the meeting. William O. McClure agreed to act as rapporteur for the symposium.

In preparing this proceedings volume, many people assisted in a variety of ways. First and foremost was A.V. Precup, Editorial Consultant on contract to the American Institute of Biological Sciences. Without her promptness and help, this volume would have been greatly delayed or might never have appeared at all. Her knowledge of grammatical and editorial matters is excellent; her competence is evident in the high quality of the completed manuscript. To her, I extend my special thanks in appreciation for her superb assistance.

Arthur Herschman, Elizabeth Zeutschel, Kathryn Wolff, Joellen Fritsche, and other AAAS staff members patiently assisted in a variety of ways. Finally, Donna Dennison at the Office of Naval Research headquarters in Arlington, Virginia, and Deborah Bowman at the ONR Detachment at NSTL Station, Mississippi, helped with a wide range of preparations and problems. My sincere gratitude to all for the help and ideas they provided, and a special thanks to Deborah Bowman, who has to put up with me on a day-to-day basis.

Bernard J. Zahuranec

Bernard J. Zahuranec,
H. David Baldridge, Jr.

1. Shark Research and the United States Navy

Most of our scientific knowledge of sharks and shark behavior has been acquired during the last four decades, largely from research supported almost exclusively by the United States Navy. Basic research on shark biology and behavior in the recent past has been conducted primarily by civilian scientists in association with the Office of Naval Research. Applied research, directed more toward the immediate needs of the military, has been largely the province of various Navy laboratories: chiefly, the Naval Ocean Systems Center, San Diego, California; the Naval Research Laboratory, Washington, D.C.; and the Naval Medical Research Institute, Bethesda, Maryland.

Justifications for the Navy's intensive support of shark research were generally twofold. Because of the real and the imagined dangers that sharks pose to humans in the sea, there was a need for applied research to develop means for countering shark aggression. Basic research on sharks was, of course, essential for an enlightened approach to solution of applied problems. Second, a fundamental knowledge of the shark's biology and sensory capabilities could provide insight into various other physiological processes, including some human ones. Progress, both applied and basic, has been very satisfying over the 40 years or so, particularly in view of the relatively recent genesis of shark science.

World War II Research

Essentially, shark research, atom bombs, television, and modern rocketry are all siblings, born of the urgencies of World War II. In fact, advanced considerations of nuclear fission, microwave transmissions, and space travel were already being nurtured at about the same time that shark authority E.W. Gudger finally became convinced that shark attacks actually occurred in waters off the United States. Gudger, then Associate Curator of Fishes at the American Museum of Natural History, titled his 1937 paper "Will Sharks Attack Human Beings?" and reluctantly conceded--on the basis of irrefutable evidence of recent events--that, "while attacks in our waters are rare, still, here and elsewhere, SHARKS SOMETIMES DO ATTACK HUMAN BEINGS." Only a decade earlier the famed American naturalist and underwater explorer, William Beebe (1926), saw sharks as "indolent, awkward, chinless cowards ...; a ladyfish has a thousand times less weight and double their courage." He scoffed at stories of shark attacks and vociferously professed that sharks posed no threat to uninjured, healthy humans in the water.

Thus, the Navy's program of serious research on sharks received its spark of life in the early 1940s not from a burning curiosity on the part of the scientific community but because some wartime servicemen were being attacked, and at times devoured, by sharks. Moreover, other servicemen knew about it and worried about it. Shark attack and, more to the point, the fear of shark attack became tactical wartime military problems, which required immediate attention (see discussion by Llano, 1963).

How serious the problem of shark attack really was during World War II has never been unequivocally established. Llano (1955), in his classic analysis of wartime sea survival experiences, found that, of the 2500-odd accounts examined, only 38 mentioned actual involvement with sharks, and only 12 of those reported casualties or injuries from sharks. However, he pointed out that his estimate of the problem was surely conservative for obvious reasons: (a) to boost morale, there was a tendency to downplay sharks as a possible cause of death; (b) the records were spotty at best; and (c) any completely successful shark attack would leave no evidence--only a notation such as "lost at sea."

In any case, during World War II, particularly as air and naval action intensified in the tropical western Pacific Ocean, it became increasingly clear that the military had a shark attack problem or, more to the point, a "fear of sharks" problem. Dread of attack had become a morale concern, which had grown in seriousness far beyond that justified by actual loss of human life to sharks or even by the true incidence of nonfatal attacks, especially in comparison to numerous other commonly accepted hazards of battle.

Thus, the first suggestions concerning the urgent need for and possible development of an effective shark repellent came in March 1942 from Harold J. Coolidge (formerly with the

Harvard Museum of Comparative Zoology) of the Office of Strategic Services. He was so stirred by the tragic story of a survivor of an aircraft accident in tropical waters that he began consulting with numerous scientific associates on the possibility of developing a chemical repellent to protect survivors in shark-infested waters. Eager endorsement of the idea by the Navy's Bureau of Aeronautics quickly led to an intensive experimental program, directed by the Committee on Medical Research of the Office of Scientific Research and Development. A number of scientists from a variety of academic, military, and private laboratories were involved. After only about a year of searching for materials having measurable repellent activity, they settled on a compound of copper acetate and a nigrosine dye to comprise a standard shark-repellent packet called Shark Chaser. (Details of the development of Shark Chaser have been extensively documented; see, e.g., Llano, 1963; Tuve, 1963; and Nelson, chapter 2 of this volume.)

Postwar retrospection in consort with discouraging results of numerous additional carefully controlled tests cast such doubt on the efficacy of Shark Chaser that it was ultimately (in the mid-1970s) phased out of the military procurement schedule. Was it worthless? Given the wartime exigencies and the immediate need for something that would protect men in the water from sharks (and the fear of sharks), how effective must a promising chemical be to justify producing it in quantity for the fleet? The consensus of wartime opinion was that a substance that was effective more than two times out of three (i.e., 67% effective), when tested on hungry, feeding sharks, would be suitable. Opinions in 1943 ranged from "more than half the time" to "nearly always effective," with one Navy officer (himself a potential user) strongly advocating that any acceptable repellent should work "almost 100% of the time." No one held out for absolutely total effectiveness, perhaps realizing that such a goal probably would be unattainable, particularly on such short notice.

Thus, the wartime repellent packet was never expected to be completely effective in preventing attack. When the crash project was undertaken in 1942, it was clearly stated that, "whereas loss of life from shark attacks was negligible, the morale factor due to the great fear of sharks was so important that a repellent substance, even though not effective under all conditions, was urgently needed" (Burden, 1943). Without doubt, field trials that at times indicated an effectiveness of more than 90% clearly proved that Shark Chaser exceeded its expectations (but see Gruber's comments in chapter 4 and Nelson's analysis of deterrence versus repellency in chapter 2 of this volume).

Nevertheless, once the wartime emergency was over and discouraging test results were given wider publicity, Shark Chaser became increasingly less valuable in combating the fear of sharks. Confidence was eroded by an expanding consensus among servicemen that Shark Chaser was "no good," "useless at best," and "never had worked." The prophecy became self-fulfilling by the inevitable failure of Shark Chaser in what previously had been its most effective mode: as a psychological aid to the fearful human in a sea filled with dangerous sharks, be they real or imaginary.

Postwar Efforts

Therefore, in the 1950s, in a spirit of "back to the drawing board," Sidney R. Galler, head of what was then known as the Biology Branch of ONR, began to address the increasingly recognized shortcomings of Shark Chaser. With wartime pressures and priorities now relaxed, shark research was put in proper perspective, and competed for available limited funding with studies on numerous other hazards of biological origin that threaten the well-being of naval personnel. According to Galler (personal communication):

> As far back as mid-1947, shortly after the creation of the Office of Naval Research, the Navy, Marine Corps, and the Army were still sorting out the experiences of military personnel under conditions of emergency and/or extreme environmental stress during World War II. The Navy created a special division, the Human Ecology Branch, within the Office of Naval Research to study personnel operations and survival under extreme environmental stresses, e.g., arctic, desert, and tropic environments. This branch was also interested in studying the biology and ecology of ichthyosarcotoxism or ciguatera poisoning, especially in the South Pacific theater of operations, and in the impact of venomous organisms, including marine organisms (in contradistinction to poisonous), on naval personnel conducting naval operations. Early on, ONR contracted with a number of scientists throughout the United States. As the interests of the Human Ecology Branch expanded, the unit evolved in the early 1950s into the ONR Biology Branch.
>
> Also during this time, ONR's Hydrobiology Program was created to study the significance of

marine biota in the planning and implementation of naval operations anywhere in the world. One of the subprograms within the Hydrobiology Program was concerned with the protection of naval personnel against poisonous, venomous, and carnivorous marine organisms. During World Ward II in certain theaters of operation (e.g., the South Pacific, Caribbean, and South America), U.S. Naval personnel and Marines had been exposed simultaneously to all three threats. In particular, there had been significant casualties in the South Pacific from exposure to venomous marine animals (for example, the stargazer or stonefish); from eating fishes that had ingested algae containing alkaloidal materials that were toxic for mammals, including humans; and from attack by sharks.

To provide a benchmark against which to judge future progress in shark science, the Navy conducted a conference entitled "Basic Research Approaches to the Development of Shark Repellents" at Tulane University on 8-11 April 1958 to assess present knowledge and to elicit recommendations from numerous authorities. The papers presented at this conference by an impressive array of international researchers covered a broad spectrum of shark science and eventually became the volume Sharks and Survival, edited by Perry W. Gilbert (1963).

In June 1958, ONR established a Shark Research Panel in association with the American Institute of Biological Sciences. Its members would be responsible for counseling ONR in its development of a coordinated shark research program and for periodically assessing the effectiveness of this program in advancing the state of knowledge about sharks. In 31 meetings from 1958 to 1970, occasionally with presentations of scientific papers to open audiences, the Shark Research Panel, under Gilbert's direction, discharged its responsibilities admirably (Zahuranec, 1978, pp. 654-655). Attesting to the relatively primitive state of postwar knowledge on sharks was the extremely basic nature of some questions on which research would be initially supported by ONR and coordinated by the Shark Research Panel: How much of a problem do sharks really pose to humans in the water? Why do sharks attack humans? What species of sharks are most implicated in attacks against humans? And, of even greater elementary nature, how many species of sharks exist and how can they be recognized one from the other?

To provide a reliable data base for evaluating the shark attack problem, ONR established the International Shark

Attack File in 1959 in collaboration with Cornell University, the Smithsonian Institution, and the American Institute of Biological Sciences. More than ten years of data collecting produced records of widely varying detail on about 1700 shark attacks--many current or recent, some historical, and some likely legendary. Inadequate funding has resulted in only sporadic record keeping over the past decade, despite the strong recommendation by Baldridge (1974), after completing an analysis of the data then on hand, that "the time to cease collecting data is certainly not just when we are beginning to learn which data to collect."

Assessment of the rapidly advancing state of shark science was provided by a conference entitled "Current Investigations Dealing with Elasmobranch Biology," which was held from 30 January through 4 February 1966, at the Lerner Marine Laboratory, Bimini, Bahamas. Scientific papers on a wide variety of subjects resulted in another volume, Sharks, Skates, and Rays, edited by Perry W. Gilbert, Robert F. Mathewson, and David P. Rall (1967). This volume and Sharks and Survival have remained the classic sources of information about shark biology and shark-human interactions.

Throughout the 1960s, ONR sponsored an intensive and extensive search for repellents conducted by Perry Gilbert at the Lerner Marine Laboratory, Albert Tester at the University of Hawaii, and others. They investigated scores of chemicals, drugs, and natural products, primarily as feeding inhibitors, and the Shark Research Panel developed a standardized protocol for testing prospective repellents. Essentially, researchers were using an Edisonian approach involving serious recommendations from shark "authorities," hunches, wild guesses, chemical warfare agents, witch doctor concoctions, and the Eastman Chemical Catalog. Field testing (admittedly) was at times relatively uncontrolled, nonquantitative, and highly subjective in evaluation of results. Nevertheless, this work was done conscientiously and provided some valuable information, even if mostly negative.

A mathematical analysis by Baldridge (1969) tended to discourage further search for chemical repellents, concluding that impracticability was perhaps an insurmountable barrier. Any number of off-the-shelf substances provide unacceptable chemical environments for sharks, but none does it fast enough and at a sufficiently dilute concentration to warrant packaging it in a size, weight, and form as an acceptable item of survival gear.

The basic question of why sharks act aggressively towards man was soon recognized as far more complex than could be answered merely by examining their feeding drive (see Baldridge and Williams, 1969, and Nelson, chapter 2 of this volume). Clearly, a much better understanding of the behavior of these "unpredictable" animals was mandatory, which, in turn, would require careful delineation of their sensory capabilities and related behavioral cues. Consequently, through the mid-1970s, emphasis in the ONR shark research program was increasingly shifted away from repellent assays and toward sensory biology, resulting in yet another Navy-produced tome, Sensory Biology of Sharks, Skates, and Rays edited by Edward S. Hodgson and Robert F. Mathewson (1978).

Current Research Goals

A conference was held in 1974 at the U.S. Naval Postgraduate School, Monterey, California, to assess the current status of shark research and to formulate directions for future endeavors (Zahuranec, 1975). Discussion of Eugenie Clark's (1974) encouraging observations on the shark repellent properties of secretions from the teleost "Moses sole" (*Pardachirus marmoratus*) led to strong recommendations by conference participants for a thorough evaluation of the active principle, pardaxin. Some of the studies resulting from those recommendations are reported in later chapters of this volume. This conference also made a formal recommendation to the Department of Defense that Shark Chaser be declared ineffective as a shark repellent and no longer be an approved item of survival gear for servicemen. That recommendation, of course, was recognized as tantamount to a statement of need, echoing from the early days of World War II: The United States Navy now had an unsatisfied requirement for a practical, effective shark repellent.

Forty years of research on sharks has taught the Navy that meeting this need will be extremely difficult at best. Each term in this four-word requirement is subject to broad interpretation and thus demands clear definition and understanding before realistic goals for success can be set:

- Practical: The quantity of material required to elicit desired effects on shark behavior, method of application, cost, user acceptance, hazard potential for users, suitability for military applications, and numerous other factors are very important in determining practicality.
- Effective: This is a quantitative term that is sorely in need of a standardized method of measurement. Of course, we must first arrive at a consensus on exactly what effect we want a repellent to produce. Clearly, effectiveness cannot long remain a matter of individual interpretation, with different objectives measured against different yardsticks.

• Shark: The median length of sharks involved in actual documented attacks is about 2.1 meters (Baldridge, 1974). Although preliminary testing of prospective repellents might utilize small, juvenile sharks, final evaluations must involve adult sharks of reasonable sizes and varied species. Obviously, shark size must be considered with any repellent effect that depends upon dosage. A reference "Navy Standard" shark 3 meters in total length and weighing 200 kilograms was proposed by Baldridge (1968) and accepted by the Shark Research Panel for evaluating pharmacological antishark devices.

• Repellent: Shark Chaser and four decades of research have brought us to the full realization that no single "magic pill" will serve all situations conducive to shark aggression against man. The term repellent as applied to present needs must be considered as plural and synonymous with deterrent. The now-recognized variety in shark attack situations and intricacies of shark behavior dictate that repellent activity be defined and measured in relation to specific conditions and to specific shark activities for which inhibition is desired (see Nelson, chapter 2 of this volume).

The United States Navy's concern with sharks remains very practical and basic: keep them from interfering with naval operations and from harassing naval personnel. The history of shark research over the past 40 years attests to the difficulty in alleviating this concern. As is so often the case, the solution of a biological problem that initially appears relatively simple and straightforward is actually extremely complicated and difficult. In going "back to the drawing board" in its continuing search for a shark repellent, the Navy's present program reflects a conviction that only through basic research on seemingly unrelated, perhaps rather esoteric, topics can truly effective solutions to such problems be effected.

Finally, there should be a close examination of the physical chemistry, kinetics of response, and related mathematics (vis-a-vis Baldridge's 1969 analysis) of any new repellent candidate (marine natural product, or whatever) before again falling into the insidious trap of impracticability. Otherwise, Nature's marine storehouse will end up being just another Eastman Chemical Catalog for a revived Edisonian approach to repellent development. Production of a practical and effective deterrent package involves far more than just finding a substance that is extremely aversive to sharks. Baldridge's analysis dealt primarily with incapacitating agents in the classical (enveloping cloud) repellent mode. Nevertheless, the geometry of this situation with respect to effective repellent activity (i.e., large volumes of water require inordinate quantities of repellent) is the demanding factor, not pharmacology and/or physiology. These same issues would be applicable to varying degrees, regardless of the mode of action of the repellent; they must be considered, whether or not the candidate substance is an off-the-shelf chemical or a biologically active material of marine origin.

References Cited

Baldrige, H.D., Jr. 1968. A proposed reference shark. Mil. Med. 133(8):654-662.

_____. 1969. Analytic indication of the impraticability of incapacitating an attacking shark by exposure to water-borne drugs. Mil. Med. 134(12):1450-1453.

_____. 1974. Shark attack: a program of data reduction and analysis. Contrib. Mote Marine Lab. 1(2):i-x, 1-98.

Baldridge, H.D., Jr., and J. Williams. 1969. Shark attack: feeding or fighting? Mil. Med. 134(2):130-133.

Beebe, W. 1926. The Arcturus Adventure. G.P. Putman's Sons, New York, reprinted 1954.

Burden, W.D. 1943. Shark Repellent. Final report No.8, dated 1 September 1943. Committee on Medical Research of the Office of Scientific Research and Development. Contract No. OEMcmr-184.

Clark, E. 1974. The Red Sea's sharkproof fish. Natl. Geogr. 145(5):718-727.

Gilbert, P.W., ed. 1963. Sharks and Survival. D.C. Heath and Co., Boston.

Gilbert, P.W., R.F. Mathewson, and D.P. Rall, eds. 1967. Sharks, Skates, and Rays. Johns Hopkins University Press, Baltimore.

Gudger, E.W. 1937. Will sharks attack human beings? Nat. Hist. 40(1):417-418.

Hodgson, E.S., and R.F. Mathewson, eds. 1978. Sensory Biology of Sharks, Skates, and Rays. Office of Naval Research, Arlington, Va.

Llano, G.A. 1955. Airmen Against the Sea. ADTIC Publication G-104, Research Studies Institute, Maxwell Air Force Base, Alabama.

_____. 1963. Open-ocean shark attacks. Chapter 13, pp. 369-386, in P.W. Gilbert, ed. Sharks and Survival. D.C. Heath and Co., Boston.

Tuve, R.L. 1963. Development of the U.S. Navy "Shark Chaser" chemical shark repellent. Chapter 17, pp. 455-464, in P.W. Gilbert, ed. Sharks and Survival. D.C. Heath and Co., Boston.

Zahuranec, B.J., ed. 1975. Shark Research: Present Status and Future Direction. Office of Naval Research, ONR Report ACR-208, Arlington, Va.

_____. 1978. The Office of Naval Research and shark research in retrospect. Pages 647-655 in E.S. Hodgson and R.F. Mathewson, eds. Sensory Biology of Sharks, Skates, and Rays. Office of Naval Research, Arlington, Va.

Donald R. Nelson

2. Shark Attack and Repellency Research: An Overview

The purpose of this symposium has been to reexamine the chemical approach to shark repellency in the light of recent discoveries of "biologically active" natural repellent substances, e.g., secretions of the Red Sea flatfish (Pardachirus marmoratus) and certain marine invertebrates (Bakus, chapter 10 of this volume; Clark, 1974, and chapter 6 of this volume; Clark and George, 1979; Primor et al., 1978, and chapters 7 and 9 of this volume; Zlotkin and Barenholz, chapter 8 of this volume). Previous attempts at finding a satisfactory chemical repellent were disappointing, and the chemical approach has been in disfavor for some time (Baldridge, 1976; Gilbert, 1976). The whole field of shark repellency up to 1960 was thoroughly treated in the book Sharks and Survival (Gilbert, 1963a), including the history of the World War II-developed chemical packet called "Shark Chaser" (Tuve, 1963). Subsequent tests for repellency were conducted on a wide variety of materials, including copper acetate, the active ingredient in Shark Chaser. Results were largely negative or inconclusive, and no really satisfactory material was discovered (Gilbert and Springer, 1963; Tester, 1963). At a U.S. Navy-sponsored meeting in 1974, researchers recommended that Shark Chaser no longer be purchased for fleet use (Zahuranec, 1975), and it was subsequently withdrawn from service. Since then, there has been no research and development program aimed specifically at finding an effective chemical shark repellent.

Using a mathematical analysis of the dispersive action of waterborne drugs, Baldridge (1969a, b, 1974, 1976) pointed out the complete impracticality of stopping shark attack by surrounding the potential victim with a cloud of toxin or other physiologically incapacitating material. Although pessimistic about the waterborne-chemical approach in general, Baldridge did allow the possibility that someday a sufficiently potent, rapid-acting, true repellent substance might be found. Few of the previous tests, however, involved the kinds of natural biological repellents discussed at the Toronto symposium. It now appears that such substances or their synthetic "mimics" hold promise, and should be investigated further.

Sharks attack people under a wide variety of circumstances, some of which lend themselves to chemical protection, and some of which do not. In some cases, chemicals may be best used in combination with other deterrent measures. To be worthwhile, therefore, a given repellent substance does not have to be 100% effective in all attack situations. This paper examines known types of shark attacks and attack situations, discusses the potential of chemical and other means of protecting individual persons, and makes recommendations for the field testing of candidate repellent substances.

Shark Attack Behaviors

Most sharks, including those dangerous to man, are generalized predators that usually feed on prey much smaller than themselves, taken whole or bitten in two. However, sharks are also well adapted to take advantage of unusual feeding opportunities, e.g., injured or dead larger animals, which they can bite or gouge pieces. Indeed, their dentition appears better suited to this opportunistic feeding mode than it does to their more common mode of swallowing small prey whole. It is this dentition, and its associated piece-feeding mode, that makes a number of sharks potentially dangerous to humans. With its large mouth and razor-sharp teeth, even a shark smaller than a person can inflict devastating wounds. Potential victims, therefore, should minimize the emission of distress signals (e.g., odors, sounds, erratic movements) which may stimulate this kind of opportunistic predation. This is particularly important when several sharks are present because the competitive effect could, if unchecked, lead to a highly dangerous feeding frenzy. Certain sharks, however, can use their dental weapons not only for food-motivated (predatory) attacks, but also for nonfeeding-motivated attacks of a defensive (antipredatory) or social nature.

Much of what we know about shark attack on humans comes from the International Shark Attack File, a collection of case histories obtained from eyewitness accounts, newspaper articles, and other reports. The file was initiated in 1958 by the Shark Research Panel of the American Institute of Biological Sciences (Gilbert, 1963b)--with funds provided by the Office of Naval Research--and an early analysis of the data was published by Schultz (1963). Unfortunately, the funding ended in 1967, so entries beyond that time were greatly reduced. Baldridge (1973, 1974) conducted a thorough examination of the data based on 1652 cases, 1165 of which were complete enough for coding for computer analysis. Although the file is the largest body of shark attack statistics available, it is far from complete--in terms of both the details of each reported case and the percentage of actual cases reported. For example, only 10% of the reported cases contain information from victims or other eyewitnesses. In Baldridge's own words (1974), "It must be conceded that technically the Shark Attack File is made up largely of hearsay evidence, mostly documented long after the event happened. But, like it or not, this is all we had to work with."

In recent years, interest in shark attack analysis has continued and, although no coordinated worldwide reporting program exists, there have been several significant contributions dealing with particular attack situations in specific geographic areas. For instance, Miller and Collier (1981) discussed 47 unprovoked attacks in California and Oregon, most of these by great white sharks (<u>Carcharodon</u> <u>carcharias</u>). Their information is relatively complete and accurate, with firsthand interviews of 87% of the victims. Randall and Helfman (1973) covered attacks by the blacktip reef shark (<u>Carcharhinus</u> <u>melanopterus</u>), a small Indo-Pacific species which attacks persons wading in extremely shallow water, such as over the upper reef flats of coral atolls. Bagnis (1968) discussed 10 attacks in French Polynesia, most (perhaps all) by the gray reef shark (<u>Carcharhinus</u> <u>amblyrhynchos</u> [= <u>menisorrah</u>]), and Wallett (1978) reported on shark attacks on the beaches of South Africa.

A recent Navy-sponsored report (Zahuranec, 1975) recommended "the establishment and maintenance of an effective reporting system for shark incidents, perhaps within the framework of a reestablished Shark Attack File." Despite a variety of attempts, as of this writing, this has not been successfully accomplished. As emphasized in the report,

> An effective reporting procedure is seen as mandatory (a) because of the need to identify possible causative and predictive factors of attack on man, (b) to provide an enlightened basis for development of anti-shark measures, (c) to identify changing trends in location and patterns of attacks, and (d) to provide information of such nature and accuracy as to allay unnecessary fears about shark attack.

One important point that emerged from analysis of the Shark Attack File was that not all attacks are feeding-motivated, as was once more or less assumed. Baldridge and Williams (1969)--in their paper "Shark Attack: Feeding or Fighting?"--pointed out that many victims bore "slash" wounds that did not seem consistent with an attempt to remove flesh. They concluded that as many as 50 to 75% of the reported cases from the file could have been nonfeeding-motivated, perhaps the "results of aggressive behavior directed at victims in an attitude of fighting rather than feeding."

Support for this view is provided by the recent underwater studies of the gray reef shark, an aggressive Indo-Pacific reef species known to attack divers. The attacks are usually sudden, high-speed strikes prefaced by a characteristic behavior consisting of tense, laterally exaggerated swimming accompanied by lifting of the snout, lowering of the pectoral fins, and arching of the back. First studied by Johnson and Nelson (1973), such exaggerated swimming is a threat display (agonistic display) and can be released by the actions of divers toward the shark, e.g., approaching, pursuing, or cornering it. The display is very conspicuous because it is exactly the opposite of what would be expected in a feeding-motivated attack; no predator forewarns prey of its intent to attack. As first demonstrated by W. Starck (Starck and Anderson, 1979) and later in experiments by R.R. Johnson, J. McKibben, G. Pittenger, and me (Nelson, 1981), displaying gray reef sharks will also attack small diver submersibles that approach and pursue them. Certainly such attacks are not predatory (attempts to eat the sub); the real question is whether they are antipredatory (shark regarding sub as a predator) or social (shark regarding sub as a competitor) in a context of dominance or territoriality. Further discussions of the gray reef shark display have been provided by Barlow (1974), Klimley (1974), Johnson (1978), and Nelson and Johnson (1980).

Therefore, shark attack can involve either of two general motivational categories: (a) feeding, in which the shark attempts to obtain food, either by swallowing the victim whole or in pieces or by biting pieces from the victim (if successful, such attacks result in the removal of flesh); and (b) nonfeeding, in which the shark regards the person, either his presence or actions, as a threat to itself or its resources, and attempts to remove that threat. The following definitions should clarify any confusion which in the past has resulted from inconsistent usage of words, such as threat, attack, provoked, and unprovoked:

- <u>Attack</u>: A relatively forceful physical contact by a shark--usually in the form of a bite, slash, or hit--directed at the victim or the gear he is wearing. It sometimes begins with a direct, high-speed approach signifying that the shark's apparent intent is attack.

(a) A <u>bite</u> involves both jaws and the forceful closure of the mouth, often with deep cutting or flesh removal. Most dangerous sharks are capable of "gouging" to remove flesh in which powerful side-to-side shaking of head and body aids the cutting action of the sharp lateral edges of the teeth. Such gouging is not common when sharks take small prey whole. (Details of the biting mechanisms of sharks are described by Moss, 1977, 1981.)

(b) As described by Baldridge (1974), a <u>slash</u> involves "open-mouthed use of the teeth, as distinguished from a bite produced by closing of the jaws." Often such slashes result from the use of just one jaw, "without matching puncture wounds corresponding to the position of the shark's other jaw."

(c) A <u>hit</u> is a forceful strike (presumably closed-mouth) by the shark's snout, as has been described between captive bonnethead sharks (Myrberg and Gruber, 1974) and has also apparently occurred in some gray reef shark attacks (Nelson, 1981).

One may speak of a <u>repelled attack</u>, in which the potential victim thwarts an apparent attack by some defensive action, including the use of weapon or repellent, or an <u>aborted attack</u>, in which the shark terminates the attack prior to contacting the victim, in the absence of any defensive action by the person.

• <u>Distress stimuli</u>: Emissions from wounded or otherwise distressed prey, which indicate an easy meal to predators. Obvious distress stimuli are (a) food odors, such as blood and tissue fluids; (b) low-frequency, pulsed sounds, such as produced by erratic struggling movements; (c) visual signals such as the "flashing" sides of a fish swimming erratically or the loss of countershading of a fish swimming off balance; and (d) electrical stimuli, that is, the bioelectric fields produced at the wound sites, detectable only at very close range. Lack of defensive or escape responses by a person bumped by a shark may also suggest that the victim is helpless or dead.

• <u>Feeding frenzy</u>: A highly excited, competitive, group-feeding pattern, usually in the presence of strong opportunistic feeding stimuli (sounds, odors). Competition is important here, and such behavior is not seen in single sharks.

• Investigation: (a) An investigative approach is swimming oriented to a person, including close circling, but without any physical contact; the shark's apparent immediate intent is investigation, not attack. (b) Investigative contact is nonforceful bumping, nudging, or gentle mouthing of the person by the shark, said to be for "testing" an object to assess its suitability as food. As defined above, attack excludes relatively nonforceful investigatory approaches and contacts, although these behaviors can lead directly to attack. Thus, investigation is an "appetitive" behavior in ethological terms, and presumably changes to attack when the shark is stimulated to the more "consummatory" actions of attempted food intake or aggression. Although an investigation/attack dichotomy is undoubtedly real (and useful for discussion), it is obviously impossible in many cases to determine the shark's intent. For example, if a shark approaches excitedly and is killed by a diver with an explosive powerhead, was its movement an attack or an investigative approach? In other cases the intent is clearer. When a threat-displaying gray reef shark breaks display and launches its maximum-speed approach, that is an attack. However, when a bait-attracted blue shark slowly circles and nudges a diver's leg, that is an investigation, until the shark opens its mouth and takes a forceful bite.

• Provocation: Any action of a person performed in some offensive, aggressive manner against the shark. The Shark Attack File separates attacks into provoked and unprovoked categories (Schultz, 1963), but how this distinction was made is not sufficiently clear. Provocation should not include all types of attack stimulation; strong food odors, for example, do not provoke a shark to attack; they stimulate it. There are, however, two degrees of provocation: (a) Contact provocation, which occurs when a person grabs, restrains, hits, spears, hooks, or otherwise aggressively contacts a shark (mild contacts, such as when a diver gently fends off a passing shark, may or may not provoke it); and (b) noncontact provocation, which occurs when a shark is chased, cornered, trapped, or exposed to sudden unusual stimuli, such as intense sounds or light flashes, or "passive" provocations, such as one's mere presence in a shark's territory (if such exists since territoriality has not yet been demonstrated in sharks).

• Threat display: A warning to the person to desist (leave or stop doing something) or suffer the consequences of an attack, best exemplified by the exaggerated swimming display of the gray reef shark which, if unheeded, can lead to high-speed attack. Such threats and attacks are nonfeeding-motivated, and can be triggered by provocations such as chasing or cornering the shark.

General Attack Situations

All shark attacks do not occur under the same general circumstances; there are several distinctly different situations in which most attacks occur, which demand different solutions to the problems of protecting people. The following four general attack situations are based on degrees of provocation or distress stimulation. They are not meant to be discrete, mutually exclusive categories, and not all attacks can be easily fit into one type, but they do serve to emphasize the diversity of attacks and the attendant problems in developing countermeasures for each.

• Situation 1. No apparent provocation or distress stimulation: Attacks in this category are, by definition, unprovoked and on "healthy" persons not emitting any obvious distress stimuli. The presumed attack motivation is feeding, with the shark regarding the victim as relatively healthy, "normal" prey. Therefore, one would expect the attackers to be large sharks accustomed to feeding on prey in the human size range.

Examples include (a) great white sharks (*Carcharodon carcharias*) in northern California attacking skin divers and surfers, perhaps mistaking the dark-suited swimmers for seals (McCosker, 1981; Miller and Collier, 1980) (In some of these attacks, the shark seemed to be stalking its prey, i.e., coming in from behind.); (b) tiger shark (*Galeocerdo cuvieri*) attacking a surfer on an air mattress off the coast of Oahu, Hawaii (Baldridge, 1974); and (c) bull sharks (*Carcharhinus leucas*) attacking bathers at South African ocean beaches, and in some rivers and estuaries (Wallett, 1978).

• Situation 2. Distress stimuli present: These attacks are also unprovoked and food-motivated, but the shark is responding to odors, sounds, etc., that indicate distress and a possible easy meal. The opportunistic feeding pattern is thus released, and even relatively small sharks may attempt to bite pieces from the victim. The person need not be the actual source of the distress stimuli; people have been attacked just by being near attractive odors or sounds, e.g., while spearfishing.

Examples include pelagic sharks such as oceanic whitetips (Carcharhinus longimanus) attacking survivors of air/sea disasters. Persons awaiting rescue are frequently injured, bleeding, weak, and unable to defend themselves (Case histories of this type during World War II are discussed by Llano, 1963). Blue sharks (Prionace glauca) have bitten divers in the pelagic waters off southern California. The divers deliberately baited in large numbers of sharks for the purpose of underwater photography; then--entering the water in the midst of strong bait odors--they were not able to watch all the sharks at once (Hall, 1981; Waterman, 1979). Similarly, gray reef sharks (Carcharhinus amblyrhynchos) have attacked divers during spearfishing activities. These tropical Pacific sharks can become highly aroused by the struggling sounds and odors of the speared fish, and sometimes mistake the diver for the source (Harding, 1975; Hobson et al., 1961; Johnson, 1978; Nelson and Johnson, 1980; Taylor, 1981a, b). Finally, blackfin (blacktip) reef sharks (C. melanopterus) attack the legs and feet of persons wading over shallow reef flats of Indo-Pacific coral atolls. These small, active sharks, unable to see the entire person, probably mistake the splashing feet for a struggling fish (Randall and Helfman, 1973).

Of course, what appears to be distressed or undistressed to a shark is debatable. A human may never appear completely "normal" when compared to other marine animals. Perhaps even an accomplished swimmer or skin diver may look rather clumsy to a white shark--more like a weak or distressed seal than a healthy one. How does a shark regard an uninjured human floating in a life jacket, or a pair of legs dangling down from a surf board? Thus, the distinction between the first two situations is somewhat fuzzy, but certainly the probability of a feeding attack increases as the apparent distress of the victim increases.

• Situation 3: Noncontact provocation: In this situation, the attack victim, perhaps inadvertently, provokes the shark by doing something that the shark perceives either as a danger to itself or in competition for its resources (Nelson, 1981). The shark's response would be called antipredatory if it regarded the person as a predatory threat, i.e., an attempt to capture the shark. If the issue is competition, one possibility is that the shark is expressing a general dominance, as suggested by Johnson (1978). Another possibility is territoriality, if the shark is defending a space and the person is regarded as an intruder, i.e., a competitor for resources in that space. Territoriality in sharks, although suggested (McNair, 1975), remains a theory without conclusive evidence. In either case, nonfeeding attacks of these types are sudden, direct strikes, prefaced by some kind of threat display.

Examples include gray reef sharks (C. amblyrhynchos) attacking skin divers (or small submersibles) that pursue and/or corner them. The high-speed strikes are almost always preceded by some degree of exaggerated swimming display (Johnson and Nelson, 1973; Nelson, 1981; Starck and Anderson, 1979). Similarly, the Atlantic species of lemon shark (Negaprion brevirostris) charge and bite after being chased by boats or having harpoons thrown at them, even when the harpoons miss (Randall, 1963). The Pacific species (N. acutidens) also has a reputation of violent retaliatory attacks in response to attempts to spear or harpoon it (Johnson, 1978).

• Situation 4. Contact provocation: Attacks in this case result from direct aggressive contact by the person, and are thus antipredatory (in self-defense) by the shark. A diver can expect to be bitten after grabbing a shark and attempting to restrain it; this happens with many other animals as well. Other contact provocations such as hitting, prodding, or spearing can also provoke retaliation, even though no attempt is made to restrain the shark.

Examples include nurse sharks (Ginglymostoma cirratum) biting divers who have grabbed their tails, prodded them, or accidentally kicked them (Randall, 1963). Often the biting shark holds on so tenaciously that it has to be killed before its jaws can be pried apart. Also, lemon sharks (N. brevirostris) have attacked after being harpooned or speared. Randall (1963) related an attack in a seawater pen at Bimini, Bahamas: Francis Blanc entered the water to spear a one-meter-long shark. "It was struck several times, but the spear did not engage the flesh. The shark then bit Blanc on the calf of the leg." In another case a gray reef shark (C. amblyrhynchos) bit a Polynesian fisherman trying to remove it from a fish trap (wier). His arm was so severely bitten that it had to be amputated (Bagnis, 1968).

Many needless attacks of this type could have been prevented simply by using good "shark sense," e.g., knowledge that a shark may bite if its tail is pulled. What is needed here is dissemination of proper advice, rather than development of protective hardware.

Existing Antishark Measures for Individuals

There are numerous methods to protect persons in the water from shark attack (see Baldridge, 1974; Baldridge and Johnson, 1969; Ellis, 1975; Gilbert and Gilbert, 1973;

Johnson, 1978; and Zahuranec, 1975). The devices discussed here are for the protection of individuals, and do not necessarily apply to area protection, for which fences, electrical barriers, and shark-fishing operations are employed.

Most of the following devices have some degree of effectiveness in the situations for which they were designed, but none approaches the ideal "universal repellent." Many, however, have not been adequately tested under sufficiently varied natural conditions, nor with enough different species of sharks. Each also has shortcomings in terms of cost, bulk, convenience, and sometimes potential danger to the user, which must be balanced against its antishark effectiveness in determining its practicality in any given situation.

Repellents

Broadly speaking, a shark repellent is any chemical, electrical, acoustical, or visual stimulus that stops a shark from approaching and/or biting a person. Repellency refers to the behavioral act of turning away or withdrawing, and has nothing to do with whether the animal is physically harmed or killed.

The standard example of a chemical repellent is "Shark Chaser," the packet of copper acetate and black nigrosine dye issued, until recently, by the U.S. Navy and Air Force. Shark Chaser and its "chemically active" ingredient (copper acetate) was discussed in chapter 1 (and later in this chapter), so suffice it to say here that it does not work adequately as a chemical repellent, and what effectiveness it may have had was due primarily to the visual screening and/or repellent effect of the black dye. No other chemical shark repellent is available, although there are products available to repel other animals, e.g., insects, fish, dogs, and humans.

There have been various attempts to develop an electrical shark-repellent device. A somewhat successful one, the "Shark Shield," was produced by Electromagnetic Industries, Clearwater, Florida in two configurations: (a) a trawl protection unit to keep sharks from damaging the cod ends of shrimp trawl nets, and (b) a diver protection unit. A battery-powered, capacitor-discharge unit is connected to two or four woven-metal electrodes which, for the diver unit, are worn over a wet suit along the arms or legs. The units were effective in keeping sharks at least 2 meters away, but the cost, bulkiness, and "feel" of the electric discharge were disadvantages for the diver. The trawl unit, according to

Gilbert and Gilbert (1963), who tested it on captive sharks at the Mote Marine Laboratory, Sarasota, Florida, was "very effective, at a pulse frequency of two per second, in repelling four species of adult sharks known to be dangerous to man." Awbrey et al. (1979) reported tests on Pacific sharks attracted to tuna in purse seines: During times when the sacked-up tuna were being brailed into the vessel, the electrical device was able to keep circling sharks at least 3 meters from the electrodes.

Experiments by Myrberg et al. (1978) and Klimley and Myrberg (1979) demonstrated that, under certain conditions, a degree of repellency can be achieved by <u>acoustical</u> means. Sharks that were initially attracted by pulsed sounds would subsequently withdraw if the playback were suddenly changed to another sound at a higher level or with a faster rise time. However, more research is needed to determine if it is feasible to develop an acoustical repellent device of a size practical to be carried by a person in the water.

Human vocal sounds can momentarily repel some sharks, i.e., cause startle responses and temporary retreat. Eibl-Eibesfeldt and Hass (1959) and Nelson (1969b) describe how spearfishermen often turn away approaching reef sharks, e.g., bull sharks, by loud shouts and aggressive arm waving. The relative contribution of the acoustical and visual elements of this has not been determined. This procedure certainly does not work with all species under all conditions, and with some (e.g., the gray reef shark) it may even trigger an attack.

At times, an aggressive posturing can have some repellent effect, and this is probably <u>visual</u>. McNair (1975) observed gray reef sharks and silvertips (<u>C. albimarginatus</u>) and reported that "most sharks show a decided sensitivity to the face plate (eyes) of a diver." Aggressive individuals were more likely to keep a distance if faced by a diver, than if allowed to come in from behind. Hans Hass (1966) described an interesting visual device--a large umbrella, on which was painted a fierce-looking face with large eyes and prominent teeth. When opened in front of an approaching shark, it frightened it away.

A controversy has surrounded one reported visual device, the black and white striped (banded) wet suit, conceived by W. Starck. As described by Doak (1974), the suit had a dramatic repellent effect on inquisitive gray reef sharks off the Australian coast. According to Starck, the suit was derived from the warning coloration of the banded sea snakes of that region, which are said to be "virtually immune to the attack of all predators." Navy scientists C.S. Johnson and J. Fish (1974) conducted controlled tests of the suit and various patterned targets both on blue sharks off California, and on gray reef, blackfin reef, and reef whitetip sharks (Triaenodon obesus) at Enewetak, Marshall Islands. They reported entirely negative results, although neither site was inhabited by sea snakes. In reply, Starck (1974) reaffirmed the general effectiveness of the suit, stating:

> A striped suit is not a magic invisible shield, nor do sharks flee in terror from it. I have been using a striped suit consistently for a year and a half, day in and day out diving in the Tasman Sea, Coral Sea, and around New Caledonia, the New Hebrides, the Solomon Islands, and New Zealand. Other divers with me almost invariably wore ordinary black wet suits. Occasionally the results were dramatic, as described in Wade Doak's article. More often, sharks were just more wary and stayed a little further away from me than from the other divers.

Thus, any repellent effect of the visual pattern is probably dependent on the species involved, encounter situation, and perhaps the geographical location. Unfortunately, Starck has apparently not published any further disclosure of his findings, such as might include quantitative results of specific experiments.

Weapons

In cases where potential victims can see the approach of the shark, they can employ a weapon to thwart the attack. A weapon is any device which can be used to drive off, injure, incapacitate, or kill the offending shark. Lacking any suitable object, as a last resort, a person's hands or feet may be used as weapons to fend off a shark.

The shark billy is the simplest weapon--a short pole used at close range to fend off sharks. More sophisticated versions are tipped with a cluster of sharpened metal points to provide better grip against the shark's skin.

The powerhead weapon or "Bangstick" is a lethal device which on contact detonates an explosive cartridge against the shark. The firing chamber is usually on the end of a short pole, which is thrust against the shark to detonate the shell. The weapon is propelled either by hand, or with the aid of a rubber sling. A good powerhead is the most effective antishark weapon, and a well-placed shot to the head (brain) can instantly immobilize or kill a shark. A fast-load powerhead developed by R. McNair has no moving parts, self-ejects the spent shell, and has sufficient power to kill the largest dangerous sharks (McKenney, 1975). Divers who work in the presence of aggressive sharks frequently carry powerheads. Like any high-powered firearm, they are potentially dangerous to the user and must be treated with care and respect. In the unarmed mode, of course, they can be used as shark billies.

The gas-injection dart, designed primarily for skin divers (Langguth, 1972), uses a small cylinder of carbon dioxide instead of an explosive cartridge. When the sharp, hollow dart is driven into the shark's lower midsection, it releases carbon dioxide which inflates the body cavity and disables the shark by rendering it very buoyant. According to Johnson (1972), the results are "devastating to the shark, whose viscera are often forced out of its mouth." Though less potentially dangerous to the user, the gas-injection dart is also generally less effective than the powerhead. It works best on a circling shark when a clear shot at the midsection is available; it is relatively ineffective against a direct, head-on approach. Some divers have also reported difficulties in penetrating some of the tougher-skinned sharks.

The electric dart is an experimental weapon that can instantly paralyze a large shark. Developed by Navy scientists D. Baldridge and C.S. Johnson, the expendable dart (which remains on the shark) is delivered by spear and produces a pulsed electric field between the inside of the shark and the outside water. As described by Baldridge (1974):

> As the dart is driven full length into the body of the shark, a switch is closed. Current then flows from the imbedded tip thru a multiplicity of pathways in the shark's body to the surrounding sea water and hence to the base of the barrel which acts as the external electrode.

A toxin syringe weapon, developed by a group of Australian divers, is similar in principle to the dart guns used to drug-capture large land animals. A barbed hypodermic needle connected to a spring-driven syringe delivers a lethal dose of toxin to the shark. The problem with this type of weapon is speed of action. According to Baldridge (1974),

> drugs tested thus far have required up to several minutes to cause death and often produced marked agitation and hyperactivity immediately following

application. This may be an insurmountable problem, for blood circulates fairly slowly in a shark, and transport of a drug from the site of injection to wherever it acts in the shark's body may simply take too long no matter how powerful the drug is in terms of toxicity.

Armor

Another approach to the shark hazard is to armor a person's body so that the shark's teeth cannot penetrate to the skin. Other problems still remain, for example, crushing due to the force exerted by the shark's jaws, the possible snapping of bones, and the bruising caused by a high-speed hit. For the most part, however, shark-attack damage is almost entirely due to the cutting, tearing action of the teeth, and if this can be eliminated, a high degree of protection will have been achieved.

A good example is the stainless-steel chainmail suit developed by underwater photographers Ron and Valerie Taylor and Jeremiah Sullivan (Taylor, 1981a, b, c). Worn loosely over a standard foam-neoprene wet suit, the finely textured suit, complete with hood, boots, and gloves weighs about 7 kilograms, and thus the diver needs little or no additional weight. Tests with divers wearing the suit have been very successful; bait-induced bites by blue sharks, reef whitetips, and gray reef sharks did not penetrate the material, and the divers did not experience an uncomfortably high bite pressure.

C. Scott Johnson, Naval Ocean Systems Center, San Diego, has tested several nonmetallic materials as possible shark-bite armor. These included the polyamide KEVLAR, which is used in bulletproof vests, and a special "armored wet suit of layered plastic 'scales' and neoprene" (Johnson, 1976).

The major disadvantages of armored suits is their high cost and inconvenience to wear. They would be most valuable to divers who must work in areas of high shark risk, since they provide continuous protection against unseen attacks and against high-speed attacks, which are difficult to stop even with a good weapon.

Stimulus Screens

Another approach is to conceal the potential victim from the shark. If attractive stimuli such as odors, sounds, electrical fields, or visual shapes and movements do not reach a nearby shark, then it will have no reason to attack.

A screen, then, works in contrast to a repellent, which must turn away sharks that are already responding to attractive stimuli. The black dye component of Shark Chaser, for example, is believed to act as a visual screen, but also possibly as a visually repelling "object" that most sharks are reluctant to approach or enter.

The Shark Screen was developed by C. Scott Johnson as a possible replacement for Shark Chaser in survival gear issued to military personnel. Its commercial production is currently being considered by SeaCo., Inc., Kailua, Hawaii (J. Fish, personal communication). As described by Johnson (1972),

> the device is a large bag made of thin, strong, very lightweight material, with three inflatable collars at the top. When not in use, the device is folded into a small package and can be carried as an accessory to a life vest or other survival gear. In the water, the package is unfolded and one or more collars of the Shark Screen are inflated orally, an operation that requires two or three breaths for each one. The person then climbs inside and scoops water into the bag by pushing down and out on the top and completely extending the bottom. Once inside the water-filled enclosure, the occupant is completely concealed. From the outside, the underwater portion of the device is a large, solid-looking, bulky object, which conceals dangling arms and legs and retains blood or other substances likely to stimulate a shark's olfactory system.

Tests show that Shark Screen is a very effective attack deterrent in the situation for which it was developed, i.e., survival situations, where victims of air or sea disasters are awaiting rescue on the surface. It cannot be used to protect a moving swimmer or a submerged diver.

Knowledge (Shark Sense)

Certainly one of the best ways to reduce the shark hazard is to provide valid advice on the characteristics of dangerous sharks and how a person should behave in their presence. Such advice comes from basic research in areas such as feeding habits, attractive stimuli, day and night activity patterns, and behavioral interactions with humans. As stated in a Navy workshop report edited by Zahuranec (1975),

> Examples of the kinds of information that swimmer/diver personnel could use are: distinguishing features of dangerous species of sharks, recognition of distinctive threat postures which may precede attack, and knowledge of characteristic activity patterns, such as the movement of blue sharks from open sea to shallow coastal waters at dusk.

Lists of specific advice to swimmers, divers, and victims of air and sea disasters have also been presented by several authors, among them Gilbert (1963c), Nelson (1969b), Baldridge (1974), and Johnson (1978).

Persons knowledgeable about sharks and how to deal with them are said to have "shark sense" (Baldridge and Johnson, 1969; Johnson, 1978). By keeping calm and knowing what countermeasures are recommended for the species at hand, such persons are more likely to emerge unscathed from a dangerous shark encounter. Good shark sense, of course, is helpful whenever needed, including those all-too-frequent times when no specific deterrent device is available.

Which Deterrent in Which Situation

Searching for an ideal "universal" repellent that is effective under all conditions is probably unrealistic; it seems better to direct the effort at improving the several kinds of antishark methods that show promise in the different situations where a need exists. For example, in survival situations, victims of air and sea disasters generally must remain on the surface in one spot awaiting rescue, often for many hours or days. Their antishark requirements are quite different from those of an abalone diver in northern California or a scientist-diver studying corals on a tropical Pacific reef.

In developing protective methods for given attack situations, we must ask the following questions: First, is the person actively moving or stationary on the surface? A moving person, such as a swimmer, diver, or surfer cannot be protected by (a) any chemical method depending on a large "enveloping cloud" about the person, or (b) the Shark Screen, a device specifically designed for stationary survivors of air and sea disasters.

Second, can the person see approaching sharks? For persons unable to see clearly underwater, such as swimmers, surfers, and survivors of air and sea disasters, a weapon or directed repellent would be difficult to use. Normally, only divers carry antishark weapons. Others wishing to include an underwater weapon in a survival kit should also include a face mask. However, even divers frequently do not see approaching sharks as they may be engrossed in other activities, such as taking pictures or looking at objects on the bottom.

Third, are rapid-strike attacks probable? Certain attacks, such as the agonistic strike of the gray reef shark, occur so suddenly and swiftly that any deterrent method must act very quickly to protect the person. Chemical methods in general would be too slow-acting in this situation; even a powerhead weapon may or may not be brought into play in time. Once such an attack is launched, it would seem unreasonable to expect any repellent--chemical or otherwise--to stop the highly motivated shark from reaching its victim. Only armor would provide complete protection in this case.

Finally, what are the limitations on size and cost? Any question of which deterrent is best eventually comes down to a trade-off between effectiveness and practicality (cost and convenience). To be practical for survival situations, for example, an antishark device must be small enough to fit in with other survival gear, e.g., life vest, marker dye, radio beacon, etc. A bulky or heavy device, such as an armored suit or a large powerhead, would be unacceptable. For the ordinary beach swimmer, a device would have to be small, very convenient to use, and inexpensive as well. On the other hand, in some high-priority commercial, military, or scientific operations, ensuring the safety of the diver is a prime consideration, so cost and inconvenience are less limiting. If sharks are regarded as a significant hazard and the work must go on, then expensive weapons or armored suits may be justified.

The following discussions do not involve chemical deterrents because, at present, an effective one does not exist. Approaches to the development of a chemical repellent, perhaps from a natural "biologically active" repellent substance, are given in the next section and elsewhere in this volume:

1. Survival situations occur in the aftermath of air and sea disasters such as ship sinkings or aircraft ditchings, and persons often must face a long time in the water while awaiting rescue. The primary concern, of course, is to survive until help arrives, however long that may take. Sharks represent only one of several problems, the major ones including hypothermia, bleeding from injuries, need for keeping afloat, need for drinking water, and need to be

spotted by searching aircraft. Since these situations usually occur on the high seas far from land, the sharks of concern are pelagic species such as the oceanic whitetip and the blue shark, species known for their bold investigatory behavior, especially when stimulated by distress signs such as blood from injured survivors. This is one of the most critical of all shark-hazard situations, and the one for which the original Navy "Shark Chaser" was primarily developed. During World War II, numerous horror stories circulated of deaths by sharks, including a reported 60 to 80, which occurred after the sinking of the U.S.S. Indianapolis in the equatorial Pacific in 1945 (Brown, 1980). (A survey of open-ocean shark attacks is given by Llano, 1963.)

Persons in survival situations normally are floating on the surface in life vests, and thus are not in a position to see approaching sharks--at least not clearly enough to take direct offensive countermeasures. Consequently, they are completely dependent on the deterrent device (if any) to provide continuous protection from sharks. There is also a deterrent size constraint, especially for pilots, to fit an antishark device in with other survival gear; a large or heavy device is unacceptable. Cost is a factor as well, but would seem less of a constraint in high-priority operations such as military flying.

The Shark Screen is a very practical antishark device for persons awaiting rescue in such situations and has other survival values, e.g., the brightly colored collar aids searchers, and some body heat is retained by the water in the bag. One disadvantage is that it cannot be used to protect a person that is swimming, such as those that might be moving about the disaster area to aid others, or those attempting to swim to shore. A Shark-Screen-type device, of course, would be recommended only if space constraints preclude a full one-man liferaft in the survival kit.

2. Beach swimming entails a certain--usually very small--risk of shark attack, although some beaches have experienced significant problems, such as the ocean beaches of the Natal coast of South Africa (Davies, 1966; Wallett, 1978). In such areas, publicly supported shark meshing programs (fishing with gill nets) have dramatically reduced local populations of sharks and, along with them, the incidence of attacks (Davis and Wallett, 1976).

For swimmers desiring individual shark protection, there is little to offer. The typical beach swimmer is clad only in a swimsuit, without mask or goggles to see clearly underwater. A practical deterrent would have to be very convenient to use and give significant continuous protection. It would seem unrealistic to expect the average recreational swimmer to purchase any device of significant bulk and cost; if there were a serious shark hazard at a given beach, most swimmers would simply go elsewhere. In theory, an extremely convenient chemical repellent for swimmers might be developed, e.g., chemical-impregnated swimsuit, repellent lotion, etc., but whether these products would gain public acceptance is questionable. Therefore, some form of area protection is most practical for reducing the shark hazard to beach swimmers.

3. Surfing involves essentially the same risks as beach swimming, except that surfers tend to spend more time farther out from the beach and in more dangerous areas than the average swimmer. In recent years in California and Oregon, surfers (and divers) have experienced an increase in shark attacks as compared to swimmers (Miller and Collier, 1981), and the culprits are believed to be great white sharks. A surfer sitting on his board awaiting a wave, relatively far out from shore, can be an inviting target for a cruising shark.

As with swimmers, surfers would probably reject bulky or costly antishark hardware, preferring to take their own risks. Area protection is much less feasible for surfers because of the extent of the areas required for the sport. Information dissemination could be useful, i.e., identifying areas of known shark risk, such as around breeding colonies of seals, which are known prey of white sharks. Perhaps in the future a repellent could be incorporated into the surfboard.

4. Abalone diving (or spearfishing) in certain temperate areas such as northern California or southern Australia involves a risk of being bitten by a white shark, as a number of documented cases attest (Miller and Collier, 1981). Oddly enough, more free divers are hit than scuba divers; perhaps the bubble sound acts as a repellent at close range (at long range it might attract), or at least decreases a person's resemblance to a seal or other natural prey. Moreover, nearly all victims of white shark attacks are soon released by the shark, as if the diver--complete with rubber suit, lead weights, etc.,--is quickly recognized as an improper food object (i.e., a case of mistaken identity). The extent of injuries inflicted--sometimes major, sometimes minor--depends largely on the chance location of where the teeth happened to penetrate.

Commercial abalone divers spend a great deal of time

underwater; they are preoccupied with their task and don't look around much; and they generate a commotion in the water that might stimulate investigation by sharks. For this situation, an armored suit would seem the best shark protection available, especially since most attacks occur without warning. Since the divers are already wearing bulky gear, the addition of a shark-proof chainmail coverall would seem acceptable (assuming reasonable cost), and would also give added protection from substrate abrasion. An armored suit would not increase overall weight (replacing the lead belt), but might increase drag. Lacking an armored suit, the diver should have a weapon or billy, which may already be part of the gear, i.e., speargun or abalone iron.

5. Open-water diving (pelagic) has increased in popularity, particularly off southern California, where divers have been deliberately baiting large numbers of blue sharks for underwater picture taking. A number of relatively minor sharkbites have already occurred (Hall, 1981; Waterman, 1979), and the potential for more serious injury exists. People enter the water in the midst of strong bait odors, and are sometimes surrounded by more sharks than they or their buddies can keep sight of at one time. A diver can relatively easily fend off blue sharks with a simple shark billy, but one has to see it first. An example is the blue shark that bit Valerie Taylor on the leg while she was trying to handfeed another (Hall, 1981).

In this situation--or any similar one where people deliberately work among numbers of bold pelagic sharks--the diver would be best protected by an armored suit. For scuba divers that don't need to move rapidly (such as photographers), the added drag of the suit would be a minor inconvenience. An electrical repellent device would also probably keep such sharks at a distance, but it may defeat the purpose of the dive by chasing away the photographers' subjects.

6. Tropical reef diving (nonspearfishing), such as on the coral atolls of the central Pacific, entails certain differences from coolwater situations. It is a warm, clear environment, in which divers often do not wear bulky wetsuits, and would be less willing to encumber themselves routinely with an armored wetsuit (the chainmail must be worn over a foam neoprene suit so the tooth tips protruding through the chain links do not reach the skin). Free divers in particular would object to this; they enjoy being able to enter the water with minimum preparations--they just slip on the mask and fins and jump in.

The gray reef shark is a commonly encountered dangerous

shark in this environment, and knowledgeable divers often carry a powerhead weapon or a shark billy for protection against these bold, aggressive animals (McNair, 1975). The gray reef shark has the habit of making very sudden high-speed strikes, although these are usually provoked by something the diver does and are usually prefaced by an agonistic exaggerated-swimming display (Johnson, 1978; Johnson and Nelson, 1973; Nelson, 1981). In this case, specific knowledge of how to recognize the threat display and how to respond to it has great deterrent value. For cases where divers must deliberately work with threatening gray sharks, e.g., if studying or filming them, an armored suit is the best protection against high-speed attacks.

7. Tropical spearfishing usually is done in the same general environment, but the situation differs because the sharks are attracted by the struggling sounds and odors (distress stimuli) of the speared fish. Many species of large sharks of both the tropical Atlantic and Pacific are capable of rapid appearances when a large fish is speared, e.g., grouper or snapper. Sound playback experiments have confirmed the role of the acoustic sense in such attractions (Myrberg, 1978; Nelson and Johnson, 1972). Acoustic attraction can bring a shark at full speed, and if the real stimulus (the fish) is not immediately apparent, the shark may mistakenly redirect its charge at the diver--a very disconcerting experience for the uninitiated (Nelson, 1969b). Such charges usually end with the shark abruptly turning away, but in some cases the excited shark continues in and bites the diver. Several sharks may also converge on the speared fish, and a feeding frenzy can ensue in which the sharks might strike at anything. Valerie Taylor was bitten twice in this way, once on the leg, once on the face, during an episode in the Coral Sea when gray reef sharks were in a frenzy over a speared grouper. She was testing the chainmail armored suit, and escaped with a few tooth punctures in her chin where the suit did not cover her (Taylor, 1981a, b).

Unlike the casual reef explorer, the spearfisherman usually expects shark action, and is somewhat prepared for it. After spearing a fish, he continually rotates during his ascent to scan the underwater horizon for approaching sharks. If one comes, he or his buddy will fend it off with a billy of some sort, perhaps a speargun or unarmed powerhead. If the shark is persistently aggressive, they may kill it with the powerhead. Wearing an armored suit, however, would be impractial for the free-diving spearfisherman because of its weight and drag, and because such divers are continuously on the alert for sharks and therefore less likely to be hit unexpectedly.

8. Military/commercial/research diving usually carries a relatively high priority and high cost per unit time as compared to sport/recreational diving. Usually the job must be done, and considerable effort may be spent in reducing the various hazards to personnel. If there is a shark hazard, deterrent equipment would probably be selected primarily on its effectiveness, and factors of cost and convenience may be secondary. For instance, the costs of armored suits should be considered minor in operations where human lives might be endangered, e.g., working among aggressive pelagic sharks while servicing oceanographic instruments or assisting in the recovery of manned space capsules. The maximum protection in dangerous situations such as these probably consist of equipping the divers (operating in pairs) with a good armored suit and a reliable powerhead weapon, which would be used most of the time only as a billy to fend off sharks.

One special consideration in clandestine military operations is the requirement that any deterrent method not alert potential enemy personnel, which could happen if powerheads were detonated or if densely colored clouds of repellent chemical were used in shallow water.

Chemical Approach to Shark Repellency: Past and Present Rationale

Recent opinions on the potential of developing a useful chemical deterrent for sharks range from the pessimistic (Baldridge, 1969b, 1974, 1976; Gilbert, 1976) to the cautiously optimistic (Bakus, chapter 10 of this volume; Bonaventura and Bonaventura, chapter 5; Clark, 1974, and chapter 6; Gruber, 1981, and chapter 4). All agree, however, that at the present time no really effective chemical deterrent is available that is practical for use by individual persons. Because of the variety of shark-attack behaviors, a repellent that always works under all conditions may never by found. To what degree this ideal should be a requirement has itself been a point of some disagreement. For instance, Baldridge (1974), himself a naval officer, concluded that "a shark repellent that only failed once in a while would be about as valuable as a parachute which opened most of the time." On the other hand, Gilbert and Gilbert (1973) commented, "The fact that a repellent does not work against actively feeding sharks or many sharks in a feeding frenzy does not mean that it is useless. A substance which can repel a curious or circling shark has unquestionable merit."

A potential repellent, then, chemical or otherwise, is worth developing if it shows promise of providing a significant degree of protection in a particular situation where a need exists. Effectiveness under all possible conditions is an unreasonable expectation and should not be a requirement, as, for example, in the pharmaceutical industry, where many useful medicinal drugs are effective on only a few kinds of diseases, and on those diseases produce a cure less than 100% of the time. A repellent that would, for instance, help skin divers ward off circling sharks should be considered worthwhile, even though it would be of little use for swimmers unable to see underwater, or for victims of air and sea disasters floating on the surface in life jackets. For the last group, a chemical method may not even be preferred, as a Shark-Screen-type device (protective plastic bag) may be the most practical deterrent.

Shark Chaser, the original U.S. Navy "shark repellent," was intended to be primarily a chemically aversive material, with copper acetate as the active ingredient. Its final formulation consisted of a mixture of 20% copper acetate and 80% special nigrosine dye (black) compressed with a water-soluble wax binder into a 170-gram cake. It was deployed by unwrapping the cake and swirling it about to create an "enveloping cloud" of the chemical around the person. The dye was included as a possible visual screen, which also reassured the user by letting him see the extent of the material spreading out around him. As detailed by Tuve (1963) and Gilbert and Springer (1963), copper acetate was chosen largely as a result of tests begun in 1942 at the Woods Hole Oceanographic Institution on captive smooth dogfish (Mustelus canis). Both copper sulfate and extract of decomposed shark meat (high in ammonium acetate) inhibited feeding responses by the small sharks. Researchers thought, according to Tuve, that combining "both copper ion and acetate ion in one compound should yield a 'double-barreled' effect," and initial field tests off Ecuador were promising. The conclusion was that "copper acetate at rates of diffusion as low as one-tenth of a pound per hour in the water surrounding an attractive bait will protect it from shark attack and should similarly protect man from being shark food."

Additional open-water tests during 1943 and 1944 seemed to confirm the general effectiveness of the Shark Chaser formula, but suggested that most of the repellency was due to the dye rather than to the copper acetate. Indeed, in tests on frenzied sharks feeding on fish discarded by a shrimp trawler, the copper acetate alone had little, if any, effect whereas the dye/copper acetate mixture did.

Some years later, Perry W. Gilbert conducted further tests on several species of large sharks in the seawater pens of the Lerner Marine Laboratory, Bimini, Bahamas (Gilbert and

Springer, 1963). Controlled tests indicated that neither copper acetate nor extract of decomposed shark flesh had any effect as a repellent or as a "feeding inhibitor." The black dye, however, kept all species away from the baits except for the nurse shark, which did not hesitate to enter the dense cloud to feed. Thus, in spite of the apparently positive results of earlier tests, copper acetate is not a satisfactory repellent substance, and any effectiveness of Shark Chaser was primarily due to the visual effect of the opaque black dye. In any kind of wave action, however, the initial enveloping cloud of dye disperses rather quickly, leaving the person without protection. Thus, at a Navy-sponsored meeting in 1974, researchers concluded that "a mounting body of evidence has now conclusively demonstrated that Shark Chaser has no significant deterrent value against most dangerous sharks" (Zahuranec, 1975) and recommended that it no longer be purchased for Navy use. It was subsequently withdrawn from service (see chapter 1 of this volume).

The disappointing experience with Shark Chaser, however, does not mean that all chemical approaches are unworkable. Chemical repellents remain attractive, at least in theory; they would have many potential advantages over other deterrent methods, e.g., simplicity of use (no electronics or batteries), small size, low cost, and ease of manufacture and distribution. The "enveloping cloud" method of deployment has problems in regard to quantity of chemical needed, but there are other ways to utilize chemicals as shark repellents, such as directed streams, repellent-tipped billies, and impregnated suits.

Responses of Sharks to Aversive Chemicals

The literature is inconsistent in regard to terminology describing the effects of chemicals on shark behavior. In some cases, the word repellent is used loosely to mean any aversive or toxic substance used for antishark purposes. But not all antishark chemicals act specifically as repellents. The term deterrent seems better for the general case, as used by Gilbert and Gilbert (1973) and Zahuranec (1975). In considering possible chemical shark deterrents, it is useful to distinguish between several distinct modes of action, in both behavioral and physiological terms. Besides repellency, possible behavioral responses include rejection, regurgitation, feeding inhibition, irritation, aggression, and distress, incapacitation, or death. These responses are not mutually exclusive, and some substances would elicit more than one of them. Relatively little has been published on the behavioral aspects of sharks in aversive chemical fields. More research in this area is needed, as it bears directly on the problem of developing practical countermeasures to shark attack.

Behavioral Responses (Observable Motor Patterns)

Repellency is the classic concept of how a "shark repellent" should work: The repellent substance, dissolved or dispersed in the water, causes the approaching shark to turn away or withdraw. Ideally, the material would be effective at a low enough concentration for the repellency to occur at some significant distance from the concentrated source of the material. Repellency is the opposite of attraction, because its definitive characteristic is that the animal responds by moving away from the source. The initial repellency might be a "startle response," a quick acceleration (explosive glide) away, but such responses are often short-lived, and the shark might circle back almost immediately for another approach. An ideal repellent would be a stimulus so unacceptable that the shark would depart with some urgency and not return. In actual practice, of course, a repellent would have to work in the presence of whatever positive stimulus (odors, sounds, etc.) attracted the shark in the first place. In such an approach/avoidance conflict, the value of the repellent must exceed that of the attractant.

A repellent that is fast-acting, especially at a reasonable low concentration, is obviously the preferred type for protecting individuals from sharks. Instantaneous repellency, i.e., response latencies of about one second or less, must presumably be mediated through external sensory receptors and neural pathways, rather than through absorption into the circulation. Examples are the natural repellent sprays of skunks and certain insects, and the secretion of the Moses sole. Many strong irritating chemicals also have such an instantaneous effect.

Speed of action, however, is not definitive to repellency, and there are probably slow-acting repellent substances to which sharks would respond in times of minutes, hours, or even days. A slow-acting substance would have little or no value for individual protection, but might conceivably be useful for long-term area protection. A possible example is the odor of rotting shark flesh, which fishermen report causes sharks to leave the area (Springer, in Tuve, 1963). The dispersion pheromones of certain insects, and of lampreys, are probably slow-acting repellents (Shorey, 1977; Teeter, 1980).

A rough distinction can be made between distance repellents and contact repellents: The former would act at the relatively high dilutions that occur as the material spreads out from its concentrated source, whereas the latter would act only when the shark actually touches or mouths the concentrated material itself. This distinction, of course, is partly arbitrary because what really matters is the concentration of the material at the shark's sensory receptors. If a large enough quantity of a normally contact repellent were dissolved in the water, it would be effective at some distance.

Rejection (from mouth) is a common type of contact response, in which the shark initally engulfs or mouths the source of the aversive chemical (e.g., a distasteful prey object), then quickly spits it out. The response is mediated by sensors inside the mouth, and the relatively concentrated aversive substance is essentially in contact with these sensors. Materials sensed in this way also act as repellents if they cause withdrawal from the source area in addition to rejection from the mouth. For instance, if a shark mouths a bait object coated with a strongly distasteful material, it might release the object and swim away; this could be called "gustatory repellency." Or it may release the object, not swim away, and try to bite another nearby bait. Distasteful substances would not necessarily turn off the feeding drive itself, just prevent completion of ingestion because of the aversive chemical in the food. Learned associations, however, may well affect future behaviors towards the distasteful object, although no data of this type are available for sharks.

A good example of mouth-rejection is the response to the sea hare (Aplysia), a marine mollusc long known to be unacceptable food for sharks. Von Uexkull (1895) reported that sharks would quickly spit out Aplysia if one was accidentally taken into the mouth, as during excited feeding on fish baits. This has been confirmed more recently by Kinnel et al. (1979), who have done biochemical work on the specific chemicals involved. Similar rejection responses were noted by Hobson (1963) from gray reef sharks in response to pieces of giant clam (Tridacna) and other molluscs. Moreover, I noted vigorous rejections when trying to feed captive young lemon sharks with pieces of fresh remora (Echeneis naucrates) instead of the shark's usual ration of blue runner (Caranx fusus). The Red Sea Moses sole is an example of a particularly distasteful prey to sharks, due to a milky glandular secretion retained in the fish's mucus coating. In addition to immediate rejection, other repellent and distressful responses occured when captive whitetip reef sharks attempted to ingest live sole (Clark, 1974, and chapter 6 of this volume).

Regurgitation involves rejection of food material after it has been swallowed, in which case the aversive chemical is called an emetic. This is one step beyond mouth-rejection, and presumably the sensing of the emetic takes place in the stomach. The cardiac glycosides found in the tissues of monarch butterflies are known emetics for birds, and cause vomiting within minutes of ingestion (Brower et al., 1968). Sharks normally regurgitate certain undigestible food remains, and they can be stimulated to vomit by injecting the emetic Ipecac into their stomachs (R. Bray, unpublished). Other known emetics for fish include antimony potassium tartrate (tartar emetic) and arsenous acid (Jernejcic, 1969). Of course, emetic action alone, because of its delay would not be suitable as a basis for a shark repellent.

Feeding inhibition reduces or eliminates an animal's interest in ingesting food or responding to feeding stimuli. This might occur either by reducing the central feeding drive itself, or by somehow rendering the normal feeding stimuli unattractive in the presence of the inhibitor substance. Feeding inhibition is a separate concept from repellency, and an inhibitor substance would not necessarily cause the animal to turn away from the source of the chemical. Orientation is not involved, and the response would occur just as well in a uniform field of the dissolved substance, devoid of directional cues.

One kind of feeding inhibition can be achieved through appetite suppression, at least in mammals. Amphetamine drugs such as Benzedrine and Dexedrine are effective in decreasing hunger in humans, and are used in the treatment of obesity (Goodman and Gilman, 1970). They are believed to act in the brain by inhibiting the feeding center in the hypothalamus (Guyton, 1976). Whether a comparable chemical-mediated effect occurs in sharks is unknown, but certainly a possibility.

Low temperature is a known feeding inhibitor in sharks, e.g., captive lemon sharks typically stop feeding when the temperature drops below about 20°C (Clark, 1963). Feeding inhibitions have also been reported for certain carcharhinid sharks during phases of the reproductive cycle, e.g., males during courtship and females during pupping on the nursery grounds (Springer, 1967); presumably these would be based on internal chemical (hormonal) factors. (Animal anorexias in general are reviewed by Mrosovsky and Sherry, 1980.)

Gilbert and Springer (1963) described the original

tests at Woods Hole in 1942 where various chemicals were tested for feeding-inhibitor effect on captive smooth dogfish. For control trials, a standard-sized piece of mackerel was presented inside a small (10 cm on a side) wire mesh box set inside the experimental tank. Test trials were identical, except that the test material was also placed inside the mesh box along with bait. Total feeding responses were tabulated based on "hunting turns" (1 point for each), "biting the wire box" (2 points), and "picking up the box and shaking it" (5 points). Since the sharks could not directly mouth the bait or test chemical source, any response reductions noted were due to the dissolved chemical spreading out into the water from the mesh box. Strong reductions (in the point totals) were obtained from a few substances, among them malic acid, maleic acid, copper salts, and decomposed shark flesh. However, they did not mention whether any repellent (withdrawal) responses occurred, as might be expected with at least some of the chemicals tested. It is therefore not clear to what degree the reduced feeding responses were due to feeding inhibition or to a repellent action in keeping the sharks away from the mesh box.

Those authors also described Gilbert's later experiments in the shark pens at Bimini, in which the movements of larger sharks were noted in response to a bait and/or a test chemical delivered by a teflon tube so as to envelop the bait in a cloud of chemical. The substance was termed a "feeding inhibitor" if it resulted in a reduction in the number of "passes" to within 30 cm of the bait. Again, without further details of the shark's behavior, this could have been caused by either true inhibition of interest in the bait or repellence of the shark away from the area of the bait. The term "repellent" was used only in cases where the shark avoided the cloud of chemical in the absence of bait. Interestingly, neither copper acetate, nor extracts of decomposed shark flesh showed either effect in these tests.

The literature frequently mentions <u>irritation</u> as a response to various noxious chemicals. Such irritation behaviors in sharks include head-shaking, gill-flexing, eye blinks, jerky swimming, etc., all of which are presumably relatively immediate manifestations, perhaps reflex in nature, of stimulation of the free nerve endings of the common chemical sense. Tester (1963) considered the common chemical sense "an ability to perceive substances of an irritating nature." Some of the strongest animal repellents are basically irritating in nature, but the occurrence of irritation responses does not necessarily mean that repellency will also occur, especially in the presence of strong attractive stimuli. For instance, Tester found that sharks would tolerate a certain concentration of OCBM (a powerful irritant) in order to feed on bait fish, despite "head-shaking, gill-flexing, and other evidences of irritating effects from the chemical." However, if a strong irritant were in sufficient concentration, it would probably repel even in the presence of attractive food odors. In terms of practicality, the question is whether irritant concentrations high enough to repel sharks could be safely tolerated by human users.

One possible response to an aversive substance--certainly not the desired one for a repellent--is for the shark to take aggressive action against the perceived source of the material or against any nearby object. In his general model of aggression in vertebrates, Archer (1976) considered pain, especially the "sudden presentation of a painful stimulus," to be an important factor in the release of attack behaviors. Rats, for instance, can be stimulated to attack one another by electric shocks or by placing them in a cage with a heated floor (Ulrich, 1966). Moreover, sharks will bite nearby objects if sufficiently annoyed, such as being pulled by the tail or being shocked through attached electrodes. In regard to chemical-induced aggressive behaviors, there appear to be no data for sharks, but the studies of Baldridge (1969a, 1974) indicate that the initial responses to toxic drugs (either waterborne or injected) was often "excitation" or "marked agitation and hyperactivity." Whether a shark attack could actually be caused by an aversive chemical is unknown, but the possibility should at least be considered.

Finally, like all animals, sharks can be distressed, incapacitated, or killed by sufficient dosages of toxic substances. One possible antishark method would be to expose the approaching shark to a toxin or anesthetic so potent that it is quickly put out of action. Waterborne drugs of this type would enter the shark's body via the gills (some acting on the gills themselves) and be distributed by the bloodstream to the site or sites of pharmacological action. Such effects are generally systemic, e.g., causing distress, disequilibrium, convulsions, paralysis, etc., by failure of the nervous system or respiratory enzymes at the cellular level. Baldridge (1969a) found that there are generally three recognizable stages in the onset of a toxic drug's effect: excitation, marked by rapid, erratic swimming; depression, with slowed, laborious swimming, resting, and arrhythmic breathing; and incapacitation, with immobility and unresponsiveness to stimuli. The latter stage could be a narcosis (if reversible) or a prelude to death.

Toxic drugs do not necessarily stimulate the chemosensory receptors, and there may or may not be an immediate behavioral response when the shark first enters the toxicant field. Even lethal doses of some toxins do not necessarily repel. Rotenone, for instance, though toxic to fishes and sharks in certain concentrations (and thus used in fish collecting), does not prevent sharks from swimming in and out of clouds of the material to eat the dead and dying fish (Bakus, chapter 10).

One major problem with the toxic-drug approach is speed of action, i.e., the time it takes for the material to be absorbed by the gills, transported by the blood, and take effect in the tissues. Another major problem is the potential ill effects of highly toxic materials on the human user and on the entire nearby marine community.

Baldridge (1969a, b, 1974, 1976) examined the practicality of the incapacitating drug approach in an analytical study considering factors such as (a) the dilutions involved in producing a field of dissolved drug around a central source, such as a person using a chemical packet similar to Shark Chaser; (b) the concentrations and exposure times necessary to effect incapacitation in sharks; and (c) the swimming speed which might apply to a shark on a direct approach to a person. He developed a mathematical model to describe an idealized hemispherical drug field, with concentration decreasing with distance from the center. Baldridge then calculated the "quantity of drug which a potential victim would have to carry to protect himself by drugging the approaching shark." The estimate from previous experimental data (Baldridge, 1969a) was that a minimum exposure integral (integral of exposure time and drug concentration) of about 10 minutes/mg/liter would be required for incapacitation of a shark using highly potent waterborne drugs such as sodium cyanide, strychnine, nicotine, tricaine, and quinaldine. The shark was assumed to approach at 46 meters/minute, and a drug field 8.7 meters in diameter was therefore required to allow the estimated 10 seconds of total exposure time. The quantity of drug required to produce such a field was from 13 to 58 kilograms, depending on how the drug was assumed to disperse. Even more would be required to maintain the field over a reasonable period of time.

While allowing the possibility that an "extremely potent drug might someday be found," Baldridge concluded that stopping an approaching shark with known toxicants is completely impractical. To be effective, the material would have to be "several orders of magnitude more toxic than sodium cyanide in terms of both dose and speed of action. It is very likely, however, that such a drug would also be equally effective against the life of the man in the water."

Physiological Mediation (Sites of Action)

In general, chemical effects on behavior are mediated either by external sensors of the three chemoreceptor systems (olfaction, gustation, and the common chemical sense), or by actions at other sites such as the gills, central nervous system, vascular chemoreceptors, internal organs, or general body tissues. If the behavior response is very rapid (i.e., within about a second of contact with the chemical), it is presumably mediated by external sensory receptors and their neural pathways.

The chemical senses of sharks have been reviewed by Tester (1963), Hodgson and Mathewson (1978), and Kleerekoper (1978), and those of fishes in general by Hasler (1957), Kleerekoper (1969), and Hara (1971). The three senses are distinct, not only in receptor morphology and innervation, but in the role each plays in the behavior of the animal. In general, olfaction is the most sensitive, capable of detecting quite dilute solutions of certain chemicals. It functions in the detection of "meaningful" substances (e.g., food odors, specific pheromones) at relatively great distances from the source. As in all vertebrates, the receptor cells are specialized bipolar neurons whose distal branches possess hairlike cilia and whose proximal branches form the olfactory nerve. They are located in the paired nasal cavities, on the lamellae of the olfactory sacs.

Gustation is generally less sensitive than olfaction. It is primarily designed to detect and evaluate more concentrated solutions, such as from food items inside the mouth, which are actually, or nearly, in contact with the receptors themselves. The receptors are epidermal cells with microvilli connected to the brain by branches of the facial, glossopharyngeal, and vagus nerves. They occur, clumped into taste buds, in the epidermal lining of the mouth and pharynx. In some fish, e.g., catfish, they occur all over the body (Atema, 1980). Sharks have no true taste buds on the outside of the body, although the somewhat similar appearing "pit organs," which occur there, might possibly have some chemoreceptive function. But the main function now appears to be mechanoreceptive (Budker, 1938; Katsuki et al., 1969; Tester and Nelson, 1967; Zahuranec, 1975).

The role of the common chemical sense is presumably to detect certain irritating substances, in relatively high concentrations, which might actually harm the animal. The receptors are nonmedullated free nerve endings (apparently distinct from those that sense touch and pain), which communicate with the brain primarily by the trigeminal nerve.

These free nerve endings occur all over the outside body surface, but are most concentrated in the mucus membranes of the oral and nasal cavities and around the eyes.

Unquestionably the main sense involved in attractions to food odors, olfaction might be assumed to be the sense involved in detecting repellents, especially for the ideal repellent that would be effective at very low concentrations. To what extent, if any, this is true is unknown. Repellents of the noxious irritant type are believed to act on the free nerve endings of the common chemical sense (Tester, 1963). Localizing a shark's responsiveness to the nasal cavity does not necessarily indicate olfaction, as this area contains free nerve endings in addition to true olfactory receptors. It is therefore difficult to determine, without electrophysiological tests, just which chemoreceptors are responsible for a given behavioral effect.

Gustation would seem to be the sense involved in food rejection by sharks, as was reported many years ago by Von Uexkull (1895, in Kleerekoper, 1978). Mashed fish mixed with quinine was detected and approached as normally as other baits, but was vigorously spat out after it was taken into the mouth. The shark's approach was probably mediated by olfaction (sensitive to the fish odor only), but the rejection of the quinine baits was due to gustation. The possible role of the common chemical sense was apparently not considered. However, Sheldon (1909) found smooth dogfish responsive to quinine (dilute quinine hydrochloride) only in those areas bearing taste buds. In contrast, dilute hydrochloric acid elicited responses from the mouth, nostrils, and all over the body, presumably via the free nerve endings.

Rejection from the mouth does not always indicate a "noxious" material. Tester (1963) found that blacktip reef sharks, accustomed to feeding on thawed frozen squid, would reject "deodorized" squid (extracted in a series of alcohol solutions), presumably because of a "lack of taste." There are many instances of rejection of nonpreferred food items by sharks (e.g., its rejection of *Aplysia*), but to what extent these are due to a specific aversive (distasteful) substance or just to the lack of the desired taste is not known.

According to Atema (1980), the distinction between olfaction and gustation is not as easy to make in aquatic animals as it is in terrestrial species such as humans. For instance, there are certain amino acids that catfish can detect with equal sensitivity by either smell or taste receptors. Unlike sharks, catfish can also detect and locate food at some distance using their "second" gustatory system--

the taste buds distributed over the whole body surface, including the barbels. The real difference, Atema suggested, is in the types of behaviors involved. Taste appears to be limited mainly to regulating food intake, "more in a reflex sense, simply stimulating certain behavior patterns, such as pick-up, swallow, and spit-out." Smell is used for more complex processes, creating "plastic filters, through which the chemical environment is viewed, that stimulate and motivate the animal into various behavior patterns, such as home recognition, prey hunting, predator avoidance, and mate selection." In addition to recognizing typical components of food items, this includes learning of odors of home streams, as in salmon, and the formation of feeding search images, as may occur in tuna.

All of these considerations bear on the question of which senses would mediate repellency in sharks and whether any development should be aimed at any one sensory channel. No final conclusions can be offered here, but irritant-type repellents would probably act on the common chemical sense, and a relatively high concentration might be needed to cause withdrawal (and this level might also affect the human user). Distasteful materials like quinine would act on the gustatory system, but probably not be an effective repellent, if at all, until actually within the shark's mouth. If an olfactory repellent is found that is effective at low concentrations, it would probably be because it has some "special meaning" to the shark, such as the various pheromones do for many other species (alarm, dispersion, sex-attraction, etc.). In general, the olfactory system seems better suited to attraction than to repulsion.

Orientation to Repellents

Assuming a shark is "repelled" by a given aversive chemical, how does it choose its escape direction? The exact form of the withdrawal response would depend on the directional cues available and the orientation mechanisms of the shark. Three general cue situations can occur: waterborne concentration gradient detectable by the shark, waterborne repellent detectable but without gradient information, and repellent detectable only upon contact with the source object. In the first case, the shark might use a klinotactic or tropotactic mechanism to steer itself out of the chemical field.

Klinotaxis--gradient searching, using essentially "one" chemoreceptor--has been experimentally demonstrated by Mathewson and Hodgson (1972) in nurse sharks orienting positively to bait-odor corridors; it is probably widespread among sharks. Klinotaxis could be used to escape a graded repellent field, but it would seem a rather slow way to do it, involving a significant amount of back and forth searching.

Tropotaxis requires the simultaneous comparison of concentration at two chemoreceptors. It has been demonstrated recently by Peter Johnsen of the Monell Chemical Senses Center, Philadelphia, on tethered bonnethead sharks (*Sphyrna tiburo*) by using a twin-syringe apparatus to deliver food extracts to one or the other nostril. Theoretically, a shark using tropotaxis would be able to escape a repellent field more quickly, by turning more directly away from the higher concentration.

Rheotaxis--orientation to water current--is another possibility, but only in situations where the current direction can be detected, such as close to the bottom. It could not work in many situations where people might utilize repellents, e.g., at the surface in deep water. Rheotaxis has been shown to play a role in food finding by bottom-dwelling species such as lemon sharks (Mathewson and Hodgson, 1972) and nurse sharks (Kleerekoper, 1978). In the case of escape from a repellent, the shark would have to respond by moving downstream, or perhaps crosstream.

Visual association is a possible mechanism that could work in the absence of gradient or current cues. The shark would have to associate the onset of the aversive chemical with a nearby object (perceiving it as the source of the chemical), and then move away from that object using visual orientation. Consider the case of a swimmer in open water surrounded by a cloud of repellent: When an approaching shark detects the chemical, it may depart simply by swimming away from the person, which is the most obvious nearby visual object. In the case of a contact repellent such as a distasteful object, a withdrawal from it might also be based on a visual association.

A kinesis is a response possibility that does not require orientation per se. Perception of the chemical would simply cause an unoriented increase in swimming speed or rate of turning, which might result in the shark escaping the chemical field. Kineses, however, are more generally known from simpler animals (arthropods, flatworms, etc.)--e.g., an isopod's response to a humidity gradient (Fraenkel and Gunn, 1961)--and would seem less likely for sharks.

Finally, a fixed escape response is possible, as suggested by Peter Johnsen, such as a 180-degree turn and accelerated take-off. Even though orientation, per se, is not involved, such a maneuver would usually take the shark quickly out of the repellent field. However, one can conceive of situations in distorted fields or eddies where circling sharks might "depart" in the wrong direction and accidentally strike the person using the repellent.

Thus, we need to determine the kinds of withdrawal responses for the various species of dangerous sharks when they encounter repellent chemicals. This is not just academic, for certain responses might actually endanger the very person the repellent was intended to protect.

Chemical Defenses of Animals

Numerous species, both marine and terrestrial, achieve significant protection against predation by chemical means. A human endangered by a shark is in an analogous position to the prey species of such animals; it is logical, therefore, to examine the prey's natural chemical defense mechanisms for clues for ways to protect people from sharks. As stated by Gruber (1981), "A marine organism may have 'invented' an effective shark repellent during the process of evolving a protective mechanism against predation."

The literature on chemical defenses is voluminous, particularly among the arthropods, which have been reviewed by Eisner (1970) and Blum (1981). Poisonous and venomous marine animals are thoroughly treated by Halstead (1959, 1978). Broadly speaking, natural animal defensive substances can be categorized as either poisons, venoms, or repellents:

Poisons are toxic chemicals, usually of nonglandular origin, that are present in the tissues of the prey species, and are thus ingested by the predator. Behavioral responses of the predator are usually delayed somewhat (often regurgitation is the first sign), and the toxic physiological effects are usually systemic, such as on respiratory enzymes, nervous transmission, or cell-membrane transport mechanisms. Since the ingested prey individual is killed, the benefit to the prey species is to the population. The predator learns to associate its unpleasant experience with some obvious characteristic of the prey, which it remembers and avoids thereafter. For this reason, many poisonous or distasteful prey species have conspicuous warning coloration (e.g., certain tropical insects and frogs).

Venoms are toxic, glandular secretions that are injected into the enemy by structures such as spines, stings, or mouth parts. Like nonvenomous poisons, the toxic effect is primarily internal (they have less effect when applied topically), and the full physiological effect may not occur immediately. Some animals use venom injection for predatory as well as antipredatory (defensive) purposes (e.g., venomous snakes).

Repellents are glandular secretions of a strongly irritating, distasteful, or odorously repugnant nature, which are secreted to the body surface or actually ejected or sprayed at the enemy (e.g., skunks). The behavioral response of the predator is usually immediate (retreat, rejection from mouth), often in time to save the life of the individual prey being attacked. The repellent substances act on external sensory receptors, and, although some such chemicals may also possess various toxic actions, these effects are secondary to the immediate repelling effect on the predator.

Many species possess conspicuous warning (aposematic) displays designed to inform predators of the consequences of attempted predation. This mechanism depends on a learned association by the predator between the conspicuous display and the distastefulness, noxiousness, or venomousness of the prey species. Such "palatability" associations are usually learned very rapidly, sometimes as a result of a single unpleasant encounter. Warning stimuli may be visual (bright colors of insects that are unpalatable to birds), acoustical (the buzz of a rattlesnake when approached by humans), and chemical (aposematic odors of certain insects). A skunk's "low-level" odor may serve a warning function, which, along with its other warning signals (black-and-white color, raised-tail display, and foot stomping), serves to remind an experienced predator of the consequences of further approach (E. Sleeper, personal communication; R. Van Gelder, personal communication). These stimuli thus become repellent in themselves, i.e., conditioned repellents. They would not be repellent to a naive predator, such as a dog having its first experience with a skunk. Such a predator will be deterred only by the trauma of receiving the skunk's full-intensity spray, which is irritating to the membranes of the eyes and nose. On its next encounter, the warning stimuli alone may effectively repel the predator.

Can this principle be applied to the repellency of sharks? Obviously, one cannot expect to condition the entire population of sharks. However, a learned association, if it forms quickly, might be used to advantage in situations where a threatening shark makes repeated passes at its potential victim. For example, suppose a person being persistently circled by a hungry shark could, like the skunk, initially repel the shark with a large dose of powerful repellent.

Some minutes later, perhaps, the shark may recover and again approach the person. If a noxiousness association was learned, the shark may now be discouraged by a much lower concentration of the repellent. Furthermore, the effectiveness of the repellent might be further enhanced if it contained a highly odorous substance in addition to a strongly irritating material. The odorous component would not in itself be initially repellent, but might enhance the formation of the odor-irritant association.

One odor to try would be morpholine, a compound known to be detected by fishes in low concentrations. In artificial imprinting experiments, a concentration of 2×10^{-4} mg/liter alters the behavior of homing coho salmon, guiding them to their home streams for spawning (Hasler et al., 1978; Johnsen and Halser, 1980; Scholz et al., 1976). Whether the inclusion of a warning odor will enhance a shark repellent to any degree is a question that can be answered only by further experimentation.

Naturally Occurring Repellent Substances

Recent renewed interest in the chemical approach to shark repellency stems mainly from the hope of utilizing some natural biologically active repellent substance, or a synthetic imitation with a similar action. This interest is largely due to the discovery by Clark (1974) of the shark-repellent properties of the Red Sea Moses sole (Pardachirus marmoratus). Numerous other animals, both marine and terrestrial, possess chemical defenses against predation, but their effectiveness against sharks is unknown. Thus, to date, the Moses sole has received the most attention as a potential source of shark-repellent material.

As described elsewhere in this volume by Clark and others, the Moses sole is particularly aversive to some sharks due to a milky secretion, emitted by glands along its dorsal and anal fins, and retained in part in the fish's mucus coating. Powerful repellent and other distressful effects occur when captive reef whitetip sharks (Triaenodon obesus) attempted to ingest live sole. Besides immediate rejection from the shark's mouth, the responses included "violent shaking of the head," "jerking away and dashing around the tank," and "holding its mouth open for several seconds to over a minute." In other tests, a dilute solution of the secretion was "100% effective" in repelling the whitetips away from bait. These tests followed the method used by Gilbert and Springer (1963) at Bimini; i.e., the solution (1%) was injected into a teflon tube of running seawater so that the further diluted solution would perfuse the bait fish hung beneath the opening of the tube.

Analysis of the Pardachirus secretion revealed that its active component, pardaxin, was an acidic protein of 162 amino acids (Primor et al., 1978). Primor and his coworkers determined that pardaxin was toxic and lethal to both teleosts and elasmobranchs, with 1-hr LD_{50} concentrations of 5.1 μg/mL/g body weight (administered externally into 250 mL seawater) for dogfish (Squalus acanthias). They proposed that the site of toxic action is the gills, and that the pardaxin disrupts the normal osmoregulatory capability of the gill membranes, resulting eventually in lethal ionic imbalances in the body fluids. If the pardaxin is injected internally, however, a concentration of about 10 times greater is required for an LD_{50} in 1 hr, suggesting that the substance acts primarily on the outside (mucosal) surface of the gills (Primor et al., chapter 9 of this volume). In terms of observable behavioral responses, a dose of 25 μg/mL administered to the head region of restrained dogfish caused an "immediate response" of struggling, opening mouth as if to gasp, and diminished opercular rate. Although pardaxin is clearly destructive to the gill membranes, mediation of the instantaneous behavioral response would seem to require action on some sensory receptors, such as the taste buds or the free nerve endings of the oral cavity.

A few tests have been conducted on other species. Working with Primor, Bonaventura and Bonaventura (chapter 7 of this volume) reported that the Atlantic sharpnose shark (Rhizoprionodon terraenovae) reacted "instantaneously" to pardaxin with excited, rapid swimming and attempts to leap out of the water. This was followed, within minutes to tens of minutes, by "irregular opening and closing of the gill slits, increased mucous secretion, and finally disequilibrium and death."

Gruber and Zlotkin (1982, see also Gruber, 1981, and chapter 4 of this volume) are developing quantitative bioassay methods for evaluating potential repellent substances in the laboratory. They have tested pardaxin on young lemon sharks (Negaprion brevirostris) at the University of Miami. In one test, the material was injected directly into the mouth of the young shark as it fed on a fish bait. At certain concentrations of pardaxin and other chemicals, the sharks were "obviously and strongly repelled, and dashed away disoriented." One of the test substances, an industrial surfactant that Zlotkin felt would "mimic" the action of pardaxin, was about 10 times more effective than the Moses sole extract (see the next section of this chapter).

Pardaxin, therefore, is clearly aversive to sharks, and can elicit several kinds of responses including mouth rejection, repellency, respiratory distress, and eventual death. Whether this substance or some similar-acting chemical can be developed into a practical shark repellent remains to be seen. Pardaxin can, in sufficient concentration, elicit the desired rapid repellent action, but it is relatively unstable except in the less-effective freeze-dried form. Results of preliminary field tests on wild sharks have been mixed, and the concentrations required for effective repellency in open water have not yet been determined. Although pardaxin itself would probably be too costly either to collect or to synthesize, it is promising that there are readily available, inexpensive substitutes that may work as well or better.

Many other marine animals, particularly invertebrates, utilize some form of active or passive chemical defense against predation. Bakus (1968, 1973, and chapter 10 of this volume) has discussed the toxicity of various tropical reef forms, especially holothurians and sponges, which protect them from being eaten by fishes. One small sea cucumber (Holothuria difficilis) abundant in exposed locations at Enewetak, Marshall Islands, was "highly toxic to reef fish." The Cuvierian organs (tubules) from these were not only ejectable on provocation--considered a defense mechanism--but were toxic themselves.

Sobotka (1962, 1965) tested holothurin, the neurotoxin from the Cuvierian organs of the Bahamian sea cucumber (Actinopyga agassizi) on a young lemon shark. When a solution of 1 gram of the material was poured into the 1520-liter circular tank, the shark showed "immediate avoidance reactions." The eventual dilution of 1:600,000 killed the shark within 50 minutes. However, holothurin is a powerful surfactant (causes foaming in water) and, in the powdered form, extremely irritating to the nasal passages of humans themselves.

More recently, Bakus (chapter 10 of this volume) tested a crude extract of halitoxin from the Caribbean red sponge (Haliclona rubens) on two swell sharks (Cephaloscyllium ventriosum) held in a running seawater aquarium. With a turkey baster, a solution of the sponge extract was administered to the head region. This elicited a response of "active body thrashing" in 3 to 7 seconds, soon followed by "convulsive respirations." Free-swimming blue sharks, however, showed no reluctance to eat extract-saturated mackerel baits, nor did they show any response to several attempts to squirt the halitoxin solution at their snouts as they swam past the boat.

The marine ophistobranch mollusc Pleurobranchus membranaceus possesses defensive glands on its body surface. When roughly disturbed, as when mouthed by a fish, these glands produce a highly acidic secretion (pH 1.0), which caused the fish to reject the molluscs (Thompson, 1960; Thompson and Slinn, 1959). And the previously mentioned sea hare (Aplysia) is another mollusc distasteful to sharks, which will be rejected if accidentally taken into the mouth. Aplysia derives its bad taste by sequestering in its tissues certain halogenated compounds from its favorite foods--two genera of red algae. Chemists have isolated and identified some of these compounds, among them three halogenated cyclic ethers, and confirmed their "antifeedant" activity in bioassays with small teleost fish (Kinnel et al., 1979).

An entirely different type of repellency, one that is not antipredatory in function, involves specific dispersion pheromones (Shorey, 1977) presumably sensed at very low concentrations by the olfactory organs. In fishes, one example is mentioned by Teeter (1980): a dispersion substance released by "sexually mature male sea lampreys, which is avoided by other males," and which "may aid in dispersing males over the available spawning habitat during nest construction." Such pheromone-induced behaviors are intraspecific and probably not very instantaneous in action; moreover, there is no indication of a repellent pheromone in sharks. The phenomenon is of interest here as an example of a "distance repellent" acting at very low concentrations and sensed via olfaction.

In those cases where aquatic animals secrete repellent substances in response to a predator, it is not a directed stream or spray as in some terrestrial species. To eject a directed stream at a predator at some distance would be mechanically much more difficult in water than in air. However, there are many active chemical defenses in land species, particularly among the insects, amphibians, and mustelid mammals. I have already discussed the repugnant, irritating spray of skunks. Among amphibians certain toads, frogs, and salamanders are unpalatable to predators because of noxious skin secretions. Some of these emissions occur in response to a predator's threat, and some can even be ejected to some distance, as in the toad (Bufo marinus) (J. Savage, personal communication). Some tropical frogs utilize bright, warning colors to advertise their noxiousness, such as the poison-dart frogs of South America, so named because their skin secretions are used by Indians to poison the tips of blowgun darts (Myers and Daly, 1976; Myers et al., 1978). One bright-yellow species (Phyllobates terribilis) can emit sufficient poison (batrachotoxin alkaloids) to make it potentially lethal to man.

The numerous chemical defense mechanisms of arthropods are reviewed by Eisner (1970) and Blum (1981), the latter cataloguing more than 600 compounds evolved by arthropods for defensive purposes. Among insects, the bombardier beetle (Brachinus) is famous for its "explosive" emission of a spray of hot quinone aimed at its adversary. This is produced by catalytically reacting two separate secretions, hydroquinone and hydrogen peroxide, in a special chitinous capsule opening to the outside. Oxygen gas produced in the reaction drives the quinone out under pressure. Other insects with directed defensive sprays include the "stink beetles" (Eleodes) (Eisner, 1966) and certain carabid beetles, cockroaches, and the walking-stick (Anisomorpha).

Some arthropod defensive substances originate in the plants they eat. For example, the unpalatable cardiac glycosides (emetics) from milkweed plants are sequestered in the tissues of monarch butterflies (Danaus plexippus), rendering both the caterpillars and the adults inedible to birds (Brower, 1969). Other chemicals of plant origin are utilized by man as animal repellents, such as capcaicum from hot peppers, which is the active ingredient in an aerosol dog repellent.

Whether the natural defensive chemicals of terrestrial origin would be any more or less repellent to sharks than those of marine origin is unknown. However, Bonaventura and Bonaventura (chapter 5 of this volume) recently found that dispersions of lyophilized fireflies at a concentration of four flies/liter caused "instantaneous" aversive responses in sharks, eventually leading to a nonlethal paralysis. A practical shark repellent, of course, requires a relatively instantaneous behavioral response, not long-term toxic or lethal effects.

Other Repellent Substances

Although investigation of natural defensive substances may well lead to the development of a practical shark repellent, the search should not be limited to such materials. We should test (or retest) any material, natural or otherwise, that may have significant repellent properties. A wide variety of compounds have already been studied during the World War II effort (Gilbert and Springer, 1963; Tuve, 1963) and during later tests at Bimini (Baldridge, 1969a; Gilbert, 1970; Gilbert and Springer, 1963), and at Hawaii (Tester, 1963). Materials included fish poisons such as rotenone, irritants such as chlorine, systemic poisons such as sodium cyanide and strychnine nitrate, narcotics such as MS-222 and quinaldine, chemical stenches, organic acids, and chemical-warfare gases. Results of these tests were usually negative, with few of the materials showing any promise. Although many sharks were killed by the test chemicals, the desired quick repellent action did not occur at the concentrations deemed low enough to be practical.

However, most of these tests were on species and under circumstances not closely resembling the typical situations when a shark repellent would be needed. Most tests were on captive sharks affected to various degrees by the conditions of captivity. These results, therefore, should not be regarded as the final word on the usefulness of any particular chemical as a repellent under natural conditions, such as when a human swimmer in the open sea is approached and circled by a shark. A general review of chemicals tested is not included here, but a few compounds are of special or historical interest:

Copper acetate was chosen during World War II as the chemically active ingredient in Shark Chaser because of the apparent effect of copper ions (from copper sulfate) and of decomposing shark flesh (high in ammonium acetate) in reducing the feeding responses of smooth dogfish (Gilbert and Springer, 1963; Tuve, 1963). Field tests of the copper acetate itself, conducted mainly on species of *Carcharhinus* and *Sphyrna*, gave mixed results. For example, in sufficient concentration, it was effective in protecting baits. In tests off Ecuador, 0.9 kilogram of the chemical in a mesh bag (diffusion rate of 0.05-0.22 kg/hr) tied 60 centimeters above a tethered bait resulted in no bites, although control baits were taken 26 times. In similar tests off Western Australia, Whitley and Payne (1947) concluded that copper acetate (diffusing at about 28 g/hr) was effective in protecting test baits. However, off St. Augustine, Florida, an aqueous solution (no concentration given) sprayed on the water surface did little to prevent excited sharks from feeding on trash fish discarded from a shrimp trawler. Other tests indicated that copper acetate's repellent value decreased with increased activity (excitement) of the sharks.

Later tests of copper acetate on captive sharks, however, showed no significant effect (although concentrations used were not given). From tests in the shark pens at Bimini, Bahamas, Gilbert and Springer (1963) concluded "that copper acetate fails to repel or inhibit the feeding activities of the several species of sharks we have worked with at Bimini." Tester (1963), working out of the University of Hawaii, reported "no irritating effects in tests with tiger and grey sharks" and that "copper acetate does not seem to be an active repellent to sharks or other fish." In regard to feeding inhibition, Tester commented:

> If it is, in fact, a feeding inhibitor, its action is probably due to the formation of insoluble compounds on the mucous membrane of the lamellae that temporarily dull or occlude olfactory perception, as suggested by F.G. Wood, Jr. He noted that when the copper concentration in a tank at Marine Studios, Florida, rose to more than 0.4 ppm, not only the sharks but all fish in the tank stopped feeding.

What is the reason for the discrepancy between the somewhat positive test reports of the 1940s and the almost completely negative results of the 1960s? Most of the differences are probably due to differences in test conditions in regard to (a) concentration of chemical used, (b) species of shark, and (c) condition of the test subjects, that is, their general health and responsiveness and whether responses were affected by conditioned feeding habits or by competitive excitement. Thus, copper acetate probably has some shark-repellent effect at some concentration, which diminishes with decreased concentration and with increased shark excitement or motivation. This generalization undoubtedly applies to many other noxious substances as well. However, the question is whether such repellent effects are sufficient to form the basis of a practical deterrent for human use, and for copper acetate the answer seems to be no.

One irritant chemical that in some tests appeared to have promise as a repellent is the powerful lachrymator OCBM (orthochlorobenzyl malonitrile),* originally proposed by the British Admiralty. At one time, it was combined with nigrosine dye as the "Wesco" shark repellent, which Tester (1962) said was "somewhat more effective in deterring captive sharks than Shark Chaser, except when the sharks were motivated by food." Tester (1962, 1963) noted that OCBM in the solvent-dissolved form was "obviously irritating" to captive gray reef sharks and a tiger shark at concentrations of 2-4 ppm respectively:

> It produced violent gill-flexing, head-shaking, and other signs of agitation by the

* OCBM in aerosol form is one of the "tear gases" used in police and military operations. It and other antipersonnel agents are discussed further in the next section.

> sharks, after which they avoided the area of concentration. There was evidence that the irritation was produced by the particulate rather than the dissolved material (the chemical is very sparingly soluble in sea water). However, at these concentrations, it was largely ineffective in preventing the sharks from feeding on dead fish thrown into the test area after the sharks had left. Not only were they attracted back into the area, but they avidly consumed the fish despite the head-shaking, gill-flexing, and other evidences of irritating effects from the chemical.

In the dispersible block form (OCBM plus dispersing agent), it was less effective, requiring about 15 ppm to repel gray sharks but caused no response in the tiger shark at this concentration. Gilbert (1961) found no response to dispersible block OCBM from nurse, tiger, and lemon sharks in the pens at Bimini. Baldridge (1974) related British tests in the Red Sea, which concluded that OCBM produced no significant repellent effect on sharks, but it did cause "swift severe burning sensations" to divers.

Hiatt et al. (1953, 1957) tested a large number of chemicals in an effort to find a repellent substance that would effectively disperse fish schools. Using the tropical marine teleost Kuhlia sandvicensis, they tested 1266 compounds and found 362 "active" ones--those eliciting responses at concentrations of 20 ppm or less. Responses designated as "slight" included "rapid mouth movements, efforts to avoid the substance, and vertical swimming," and "violent" reactions included "very erratic and rapid swimming, leaping out of the aquarium, paralysis, head shaking, blindness and death within two or three minutes." As of the 1953 paper, the three most potent materials found were phenacyl chloride, phenacyl bromide (both powerful lachrymators: eye irritants) and allyl isothiocyanate (general skin irritant), which produced "medium responses" at a concentration of 0.1 ppm. In the 1957 report (not seen by me, but cited by Tester, 1963), they reduced the threshold for phenacyl chloride to 0.005 ppm, the lowest threshold obtained. There appears to be no published indication of whether any of these chemicals were ever tested on sharks.

In 1963 Tester reported the "surprising" find that several species of sharks showed "an aversion to human sweat at a concentration of about 1 ppm or more." Others had previously shown that human hand rinse was repellent to migrating coho salmon and that the L-isomer of serine was apparently responsible (Brett and McKinnon, 1954; Idler et al., 1956). Tester found that L-serine in concentrations above 1 ppm caused "violent head-shaking and gill-flexing in the tiger shark" but no obvious response in gray reef sharks. Steinberg (1960), however, found no repellent activity of sweat or its components against lemon sharks at Bimini. Although Tester's results warrant further investigation, he concluded that "neither human sweat nor L-serine can be regarded as a sufficiently active shark repellent to protect man for sharks."

Among other fish repellents, Hasler and Wisby (1959) patented a method for repelling fish, such as salmon fingerlings, using potassium phenyl acetate. Tester (1963), however, found no repellent effects on sharks exposed to "low concentrations" of this material; neither did Gilbert (1961) in tests at Bimini. Other research has revealed that, although not yet tried on sharks, the larvacide TFM (3-triflouromethyl-4-nitrophenol) has repellent effects on spawning-run sea lampreys (Teeter, 1980). In a two-choice preference test, lampreys avoided water containing TFM in concentrations above about 2 mg/L ppm.

Finally, Gruber and Zlotkin (1982) (see also Gruber, 1981, and chapter 4 of this volume) recently found that a common industrial surfactant (sodium lauryl sulfate) was apparently more effective in repelling lemon sharks than an extract of the Moses sole secretion. Surfactants, which are detergents capable of emulsifying oils, have long been known to be toxic to fishes, especially the lipophilic types (Parker, 1971), which attach to and destroy the phospholipid component of the gill membrane. Several natural toxic or repellent substances also possess surfactant action, such as pardaxin from the Moses sole (Zlotkin and Barenholz, chapter 8 of this volume) and holothurin from sea cucumbers (Sobotka, 1965). Industrial surfactants, therefore, show considerable promise as repellent materials, especially in view of their availability, low cost, and inoffensiveness to humans.

Development and Delivery of Repellents

Dissolved vs Particulate Material

An important consideration in the effective dilution of a repellent at the shark is whether the substance is dispersed in the form of a true solution or a suspension of solid or liquid particles. Tester (1963) observed that the irritant effect of OCBM on sharks was apparently due to the particulate rather than the dissolved material. The action of particles on the shark's sensory endings could be likened somewhat to that of a "contact" repellent. At the dimensional level of a single sensory cell, a particle of repellent could bring essentially full concentration on that cell. Thus, the cell might respond even though it has only a relatively low sensitivity to that chemical and would not respond if exposed to some moderate dilution of the fully dissolved material. The problem then involves determining how many receptor cells must be stimulated simultaneously to release a behavioral response.

Arguments both pro and con can be made for which form would make the better repellent--a true solution or a particulate suspension--and this would depend on the number of cells that must be stimulated, the individual thresholds of these cells, the size of the particles, and perhaps the length of time each particle remains in contact with a single cell. Hiatt et al. (1953) addressed this problem because many of the organic irritant chemicals they tested were very insoluble in seawater and had to be emulsified into small droplets (essentially a colloidal sol) with a dispersing agent such as Tergitol NPX. Using one powerful irritant (allyl isothiocyanate), they prepared two suspensions using the dispersant and one true solution by heating the water until the material dissolved. At an overall concentration of 0.1 ppm, the true solution evoked a "discernibly greater activity" than the suspensions, which had average droplet diameters of 3.0 and 4.5 micrometers. The response to the two suspensions was judged equal.

In the terrestrial environment, active chemical defenses involving strong irritants are more comparable to the "particulate" case. In species that emit sprays aimed at attacking predators (e.g., skunks, bombardier beetles), the repellent lands on the predator as liquid droplets at full concentration. Also, antipersonnel chemicals such as tear gases are usually delivered as particulate aerosols (solids or liquids) rather than as true vapors.

Effects on the Human User

An important consideration, of course, is the effect of the repellent chemical on the immersed human user. Obviously, many noxious irritating or toxic substances would be too harsh to consider as a practical repellent for human use. Besides the irritation of exposed body surfaces, a person could take in significant amounts of material by inadvertently swallowing the water or by absorbing them through membranes of the mouth or even through the skin. This could be a serious problem for chemicals that have some activity as systemic poisons. The concern would be greatest for any material deployed in the enveloping cloud manner (as was

Shark Chaser.) The human user, remaining at the center of the cloud of chemical, would receive a much larger dose for a longer period of time than any approaching shark that is to be repelled. Aimed delivery of the chemical via a "squirt gun" or similar device would cause less of a problem in this regard.

What is needed is a repellent whose action is relatively specific to sharks (or fish) and which has essentially no effect on humans; short of this ideal, the problem becomes a trade-off between the two. A person might tolerate some temporary irritation, but investigators would have to determine to what extent this is practical. Baldridge (1974) related the reactions of divers who were involved in repellent tests for the British Ministry of Defense in 1955:

> An aerosol preparation of OCBM was found to cause swift severe burning sensations to divers even when using protective clothing and was therefore deemed to be impractical. None of the divers was willing to test a dispersible solid block of OCBM after experiencing effects of the aerosol.

Thus, there is a real question as to whether OCBM or any other <u>anti-personnel</u> or <u>chemical-warfare</u> agent, even if effective against sharks, could ever be made suitable for use by unprotected persons in the water. This is primarily a packaging and delivery problem, i.e., how to deliver a strong dose to the shark without having much effect on the person. Chemical weapons and their effects on humans have been reviewed in a World Health Organization publication (Anon., 1970), and categorized into three general types:

- "lethal agents," used either to kill an enemy or to injure him so severely as to necessitate his evacuation and medical treatment. Examples are the "blood gases" (e.g., hydrogen cyanide), "nerve gases" (e.g., agent VX), "lung irritants or asphyxiants" (e.g., phosgene), and "vesicants or blister agents" (e.g., mustard gas).

- "incapacitating agents," used to put an enemy completely out of action for several hours or days, but with a disablement from which recovery is possible without medical aid. Examples are certain "bacterial toxins" (e.g., staphylococcal enterotoxins) and the "psychochemicals" (e.g., LSD and agent BZ).

- "harassing agents" (short-term incapacitants), capable of rapidly causing a temporary disablement that lasts for little longer than the period of exposure. Examples are the "tear gases" (e.g., OCBM or CS gas, chloroacetophenone or CN gas, and diphenylaminechlorarsine or DM gas) and some "skin irritants" (e.g., phosgene oxime).

Of these three categories, obviously only the harassing agents could be considered as possible shark repellents for human use. According to the WHO report, harassing agents

> may be distinguished from such incapacitating agents as the so-called psychochemicals in that the effects of the latter have a delayed onset and persist for a period greatly exceeding that of exposure. A distinction between these two classes of chemical warefare agent may also be drawn on physiological grounds: psychochemicals and allied incapacitating agents are centrally acting substances, while harassing agents are sensory irritants. Their harassing effects, which arise from the reflex responses of the body to sensory irritation, include lachrymation (eye irritation/ tear flow), sternutation (upper respiratory tract irritation/ sneezing, mucus flow), vomiting and pain.

<u>Methods of Repellent Delivery (to the Shark)</u>

After researchers develop a satisfactory repellent material, they must decide how it should be packaged and used. Except for the familiar "cake of soap" packaging of Shark Chaser, little has been written about this aspect. The following suggestions assume the use of a repellent chemical that has effectiveness at some distance in the dispersed form and would also have contact-repellent activity in its concentrated form:

1. <u>Chemical packet (cake)</u>: A simple packet of the Shark-Chaser type would be convenient to manufacture, distribute, carry in survival kits, and use when needed. Despite the failure of Shark Chaser itself (because its active ingredient was ineffective), the packet method of delivery remains a viable alternative. Improvements could be made in the binder matrix to give a more controlled release of the active material. Bonaventura and Bonaventura (chapter 5 of this volume) describe a hydrophilic polyurethane material they used as a slow-release matrix; they also discuss the possibility of a more sophisticated vesicle-mediated transport to deliver the active compound to specific receptor sites.

When needed, the user would open the packet by removing

a cap or peeling off a plastic sheet, etc., and possibly swirling it about to create a cloud of repellent around him. However, there are material-quantity problems with depending on such an enveloping cloud for prolonged protection, unless the repellent is an extremely potent material. For a stationary person in calm water, the enveloping cloud would be roughly hemispherical, with the user at the center and the concentration diminishing from the center to the perimeter (for formulas, see Baldrige, 1969b). In heavy wave action, the repellent field would dissipate outward more rapidly and unevenly than in calm water. If the person swam, the chemical would stream out in a corridor behind him, leaving the person with little or no repellent in most directions.

The user of a chemical packet, however, need not depend passively on the dispersed cloud of material. If he sees the approaching shark, the person can direct concentrated repellent at it by wafting the packet in front of it or even rubbing the material on the shark's snout. For the latter, some kind of extension handle on the packet would be useful. There should also be a convenient way of resealing the packet to conserve the material when it is not needed. In this regard, the black dye used in Shark Chaser was included partly to allow the user to see how well he was being enveloped by the repellent. Although that in itself was helpful, it also created the distinct disadvantage of rendering the person essentially "blind" inside the opaque cloud, inhibiting his ability to see approaching sharks or to determine whether his own legs were dangling down below the dye in plain view of the sharks.

Of course, for victims of air and sea disasters floating at the surface awaiting rescue, the Shark-Screen-type deterrent (plastic bag) has proven quite effective in tests. A chemical method may thus not be needed for this situation, or may be provided for additional protection to be used, if needed, to discourage particularly inquisitive sharks circling the bag.

2. Slow-leaching suit or strips: One disadvantage of using a single chemical packet, especially if used attached to one part of the body (e.g., the waist), is that the chemical emanates from that one point source. Other parts of the body, therefore, may be left unprotected, especially if the person is swimming and the chemical emissions are streaming backwards in a relatively narrow corridor. A shark could conceivably come in from the side or from below and bite an arm or leg without ever encountering the repellent material. For divers, this problem could theoretically be solved by impregnating or coating the entire wet suit with a slow-

leaching chemical. Repellent would then be emanating from all parts of the body, with none left unprotected. An extended enveloping cloud would still form around the diver when he is relatively stationary in the water. An alternative would be to coat the suit with a taste-aversive material only, which might last longer, but which would be generally less effective.

A modification of this approach is to use Velcro tape to attach strips of impregnated material to the suit at various key places (e.g., ankles, thighs, wrists, etc.), rather than impregnate the whole suit. Designing a satisfactory chemically impregnable diver's suit may be a challenge since ordinary closed-cell foam neoprene would probably not impregnate easily. The strip method would be simpler, but give less complete coverage. Either method would require more diver-preparation time than simply carrying a packet, but would provide continuous emissions for protection against unseen shark approaches. On the other hand, the leached-out suit or strip material would have to be replaced periodically or reimpregnated with the repellent chemical. Using both the suit and one of the shark billy devices would give the maximum protection.

3. <u>Repellent-tipped shark billy (contact applicator)</u>: For divers or other persons equipped to see clearly underwater, the simple shark billy (short pole used to fend off sharks) is generally effective. Its effectiveness could be further enhanced by the addition of a chemical repellent at the tip. A slow-release-matrix stick of repellent (much like lipstick or stick deodorant) could be mounted on the end of the billy so that it could be uncovered when needed. If a persistent shark is not discouraged by physically pushing it away, the chemical tip could be exposed and used directly against the shark's snout or mouth, delivering a concentrated dose of repellent from a relatively safe distance.

For compactness, the design should be telescoping, perhaps about one meter long, which would collapse to a fraction of that length for storage. It could be easily carried on a diver's belt, or included in survival gear (along with a face mask for seeing underwater).

4. <u>Squirt gun (underwater "Mace")</u>: This method would also be primarily for divers or other persons equipped to see underwater. Liquid repellent (concentrated solution or suspension) could be applied as a directed stream into the face of the shark, much like "Mace" or tear gas is used against human attackers. The applicator device could be a simple plastic squirt bottle, or a more elaborate syringe-type gun,

perhaps on a telescoping pole. In comparison to the solid matrix method, liquid application would allow a more powerful dose to be delivered in one shot. The repellent solution should also be slightly colored with dye so that the user could see the repellent stream. Disadvantages would include the possibility of prematurely using up the repellent supply, and the greater complexity of a squirt-gun mechanism.

5. Repellent injector (stinger): Another variation would inject the chemical into or below the shark's skin using a short hypodermic needle/syringe at the end of a shark billy. Such a "stinger," patterned after the defensive weapons of bees and wasps, would make the most efficient usage of the quantity of repellent chemical available. The appropriate chemical must be an instantaneous-acting irritant or pain-inducer (affecting the subdermal free-nerve endings), not a systemic toxin requiring distribution by the blood circulation before it takes effect. The shark's response, furthermore, would have to be one of repellency, with the shark associating the "sting" with the human user and therefore retreating from him. Such a device would simply be a chemically enhanced shark billy, and not a lethal chemical weapon, such as the toxin-syringe weapon discussed earlier, which was designed to incapacitate a large shark.

Testing Repellent Substances

For any candidate repellent material, there should be a two-phase testing program, consisting of (a) laboratory tests on captive sharks under controlled conditions and (b) field tests on dangerous species in the natural environment. Later in this volume, Gruber (chapter 4) suggests procedures for the laboratory assay of chemical effects, using standard species such as juvenile lemon sharks for which much baseline physiological and behavioral information is available. Gilbert and Springer (1963) have described the original laboratory tests on smooth dogfish at Woods Hole and the subsequent tests on larger sharks in the seawater pens at Bimini. Tuve (1963) reviewed the field tests off Ecuador, Florida, and Mississippi that were associated with the development of Shark Chaser. And Clark (chapter 6) describes both laboratory and field testing of the repellent effects of the Red Sea Moses sole. In the final analysis, of course, there must be field tests on appropriate species under appropriate conditions to determine the true effectiveness of any candidate repellent substance. The following suggestions for field tests are based on species and situations in my personal experience.

For free-ranging sharks in general, one must use bait

attraction to obtain the numbers needed for statistically significant tests. A good location for this is the pelagic environment off southern California, where large numbers of blue sharks are available most of the year, and considerable numbers of mako sharks during the summer. These sharks can be readily attracted to the drifting boat with cut mackerel in a cannister, and can be observed from a surface platform, from an underwater cage, or from an underwater viewing chamber in the research boat. The test chemical can be presented in various ways, depending on the ingenuity of the reseacher in building apparatus. Perhaps the simplest way for initial tests is to use an elongated squirt gun and apply the material directly in the face of the shark as it cruises by or as it approaches bait. As with all experiments of this type, the test trials (squirts of test chemical) must always be compared to control trials (squirts of seawater).

Researchers in a drifting boat with bait-attracted sharks could also use a version of the method described by Gilbert and Springer (1963) for the Bimini tests. The purpose is to induce a shark to approach up a bait-odor corridor in such a way that it encounters the test chemical prior to reaching the bait. The apparatus would consist of a bait cannister suspended below the boat in conjunction with a flexible plastic tube through which seawater would be gravity-fed from a topside reservoir. For control trials, only seawater would perfuse the bait; for test trials, a measured quantity of chemical would be injected with a hypodermic syringe into the upper end of the tube, so that the test solutions would flow out the tube and envelope the bait cannister.

A variation of this method would eliminate the bait cannister and use only a bait-odorized stream of water emanating from the tube (which is terminated by a perforated-ball diffuser). In this case, control trials would test the shark's response to the odorized water alone, and test trials would mix the repellent chemical with the odorized water. For either method, the observer must (a) be able to tell exactly when the shark's head enters the chemical field; (b) know the concentrations within the field (from previous measurements); and (c) between trials, allow the chemical from the previous trial to flush away (this would be no problem if the boat is wind-drifting).

Another experimental design would use a continuous corridor of bait-odor/repellent mixture (with several different repellent concentrations). The observer would determine how far up the corridor the shark swims before being repelled (i.e., how far from the perforated ball). This repulsion distance could then be related to concentration, based on the rate the corridor spreads with distance.

Several questions might be asked when designing tests for deterrent effect:

1. Response description: If the shark responds when entering the chemical field, what is the form of the response? Is it a repellency (turn away or withdrawal) or perhaps an "irritation" response without repellency (head shaking, gill flexing)? What is the take-off direction and speed? Does the shark orient to the chemical gradient, perform a "fixed escape response," or use a visual association to select the withdrawal direction?

2. Response latency, concentration threshold: How soon does the response occur after the shark enters the chemical field? Is it essentially instantaneous, or does it take several seconds or even minutes? What is the threshold concentration for repellency? Does concentration affect latency?

3. Effect of learning: After being initially repelled upon contact with the test chemical, will a shark return up the bait corridor? If so, will it show increased tolerance to the chemical (habituation) or increased sensitivity? Is there any evidence that the shark learns a warning (aposematic) odor associated with the primary noxious chemical?

To answer these questions, the observer must be able to recognize individual sharks. Blue sharks, for instance, readily tolerate being tagged for this purpose with color-coded dart or "clothespin" tags. Some will stay in the baited area for many minutes, even hours, leaving occasionally for a few minutes and then returning.

For direct human testing, persons wearing stainless-steel chainmail suits can easily withstand bites from blue sharks and certain other species (Sullivan, personal communication; Taylor, 1981a, b, personal communication). Such "voluntary" bites can be obtained by placing fish fillets under the mail of the suit, but even without bait on the person blue sharks will eventually bite someone quietly maintaining position in a bait-odor corridor. The safety thus afforded by the chainmail armored suit enables divers to test repellent-delivery systems directly under conditions simulating those of real attack situations, such as a wounded person floating in a life-jacket at sea.

As described by Tuve (1963) and Clark (chapter 6 of

this volume), tests can be designed using alternating protected and unprotected baits on a longline fishing apparatus; baits can then be checked periodically (perhaps automatically) to determine when or if they are taken. This method seems particularly applicable for tests of taste-aversive or contact repellents, for which the bait would have to be mouthed, or nearly mouthed, before a repellent dose is received by the shark.

It is usually difficult to conduct controlled field tests on nonbaited sharks because most species cannot be approached closely enough unless baited. Exceptions include the sedentary bottom-resting species such as nurse sharks, angel sharks, etc., but these are not the species for which attack deterrents are needed. Among dangerous species, the gray reef shark of the Indo-Pacific often boldly approaches divers close enough for repellent tests. Moreover, this species would present an opportunity to test repellents under motivations other than feeding (Nelson, 1981). Some gray reef shark attacks are agonistic, and this may also be their motivation for approaching divers. Certain pelagic sharks, notably the oceanic whitetip and the silky, will also approach swimmers closely on occasion, as happens around some deep-moored buoys off Hawaii and in the Tongue-of-the-Ocean in the Bahamas (Nelson, 1969a, b). Situations such as these can provide an opportunity for testing candidate repellents under natural, nonbaited conditions.

Another test approach is to determine if a shark's known daily movement patterns can be disrupted by placing a source of chemical within its home range. This type of study would apply to the problem of long-term area protection, such as ridding beaches, bays, etc., of sharks. A proposed study of this type involves biotelemetry--instrumenting home-ranging species with implanted ultrasonic transmitters, and then carefully monitoring their normal day and night patterns of activity. Chemical tests would begin after repeatable movement patterns were well-established. For instance, if a shark returns to a given specific area at a certain time each day, then a cannister emitting the test chemical could be placed there, and researchers could determine (by telemetry) whether the shark subsequently avoids the area or whether it eventually habituates, etc. If a shark typically moves from point A to point B at a certain time, a repellent corridor could be set up so the shark would have to cross it to travel its usual route; alterations from its normal route would indicate avoidance of the test chemical. In such tests, the shark would encounter the chemical field in the complete absence of any other distracting stimuli such as boats, divers, and apparatus, and very subtle responses might therefore show up which would be missed in other test situations.

A suitable species for such tests is the lemon shark in its natural shallow inshore habitat in Florida and the Bahamas. An initial telemetric study of these sharks in Bimini, Bahamas, revealed considerable day-to-day repeatability in their movements (Gruber and Nelson, in preparation). We plan to continue these studies using more automated positioning systems, including an array of data-logging monitors and a radio-relay X-Y position-plotting system (McKibben and Nelson, 1981).

References Cited

Anonymous. 1970. Health Aspects of Chemical and Biological Weapons. World Health Organization, Geneva.

Archer, J. 1976. The organization of aggression and fear in vertebrates. Pages 231-298 in P.P. Bateson and P.H. Klopfer, eds. Perspectives in Ethology, Vol. 2. Plenum Press, New York.

Atema, J. 1980. Smelling and tasting underwater. Oceanus 23(3):4-18.

Awbrey, F.T., T. Duffy, W.E. Evans, C.S. Johnson, W. Parks, and J. DeBeer. 1979. The Tuna/Porpoise Problem: Dedicated Vessel Research Program (First Leg of Third Cruise). Southwest Fisheries Center, La Jolla, CA, Adm. Report LI-79-11. 29pp.

Bagnis, R. 1968. A propos de 10 cas de blessures par requins chez des pecheurs sous-marins en Polynesie Francaise. Med. Tropicale 28:368-373.

Bakus, G.J. 1968. Defense mechanisms and ecology of some tropical holothurians. Mar. Biol. 2(1):23-32.

_____. 1973. The biology and ecology of tropical holothurians. Pages 325-367 in O.A. Jones and R. Endean, eds. Biology and Geology of Coral Reefs, Vol. 2: Biology 1. Academic Press, New York.

_____. 1981. Chemical defense mechanisms on the Great Barrier Reef, Australia. Science 211:497-499.

Baldridge, H.D., Jr. 1969a. Kinetics of onset of responses by sharks to waterborne drugs. Bull. Mar. Sci. 19(4):880-896.

_____. 1969b. Analytic indication of the impracticability of incapacitating an attacking shark by exposure to waterborne drugs. Mil. Med. 134(12):1450-1453.

_____. 1973. Shark Attack Against Man: A Program of Data Reduction and Analysis. Unpublished report to Office of Naval Research. Mote Marine Laboratory. 127pp.

_____. 1974. Shark Attack. Droke House/Hallus, Inc., Anderson, S.C. (Also Berkley Medallion Books, New York).

_____. 1976. A reminder of the impracticability of chemical shark repellents. Page 18 in W. Seaman, Jr., ed. Sharks and Man: A Perspective. Florida Sea Grant Program Report 10, Gainesville, Fla.

Baldridge, H.D., Jr., and C. S. Johnson. 1969. Antishark measures. Naval Res. Rev.. 22(4):15-22.

Baldridge, H.D., Jr., and J. Williams. 1969. Shark attack: feeding or fighting? Mil. Med. 134(2):130-133.

Barlow, G.W. 1974. Derivation of threat display in the gray reef shark. Mar. Behav. Physiol. 3:71-81.

Blum, M.S. 1981. Chemical Defenses of Arthropods. Academic Press, New York.

Brett, J.R., and D. MacKinnon. 1954. Some aspects of olfactory perception in migrating adult coho and spring salmon. J. Fish. Res. Bd. Can. 11(3):310-318.

Brower, L.P. 1969. Ecological chemistry. Sci. Am. 220(2):22-29.

Brower, L.P., W.N. Ryerson, L.L. Coppinger, and S.C. Glazier. 1968. Ecological chemistry and the palatability spectrum. Science 161:1349-1351.

Brown, C. 1980. Survivors of U.S.S. Indianapolis still relive their ordeal. Los Angeles Times August 27:9(1A).

Budker, P. 1938. Les cryptes sensorielles et les denticles cutanes des plagiostomes. Ann. Inst. Oceanogr. 18:207-288.

Clark, E. 1963. The maintenance of sharks in captivity, with a report on their instrumental conditioning. Pages 115-150 in P.W. Gilbert, ed. Sharks and Survival.

D.C. Heath and Co., Boston.

_____. 1974. The Red Sea's sharkproof fish. Natl. Geogr. 145(5):718-727.

Clark, E., and A. George. 1979. Toxic soles, *Pardachirus marmoratus* from the Red Sea and *P. pavoninus* from Japan, with notes on other species. *Environ. Biol. Fishes* 4(2):103-123.

Davies, D.H. 1966. *About Sharks and Shark Attack.* Hobbs, Dorman & Co., New York.

Davis, B., and T.S. Wallett. 1976. Anti-shark measures as practiced in Natal, Republic of South Africa. Page 16 in W. Seaman, Jr., ed. *Sharks and Man: A Perspective.* Florida Sea Grant Program Report 10, Gainesville, Fla.

Doak, W. 1974. Revolutionary weapon--anti-shark wet suit. *Skin Diver Mag.* (June):86-89.

Eibl-Eibesfeldt, I., and H. Hass. 1959. Erfahrungen mit Haien. *Z. Tierpsychol.* 16:733-746.

Eisner, T. 1966. Beetle's spray discourages predators. *Nat. Hist.* 75(2):42-47.

_____. 1970. Chemical defenses against predation in arthropods. Pages 157-217 in E. Sondheimer and J.B. Simeone, eds. *Chemical Ecology.* Academic Press, New York.

Ellis, R. 1975. *The Book of Sharks.* Grosset and Dunlap, New York.

Fraenkel, G.S., and D.L. Gunn. 1961. *The Orientation of Animals.* Dover Publications, New York.

Gilbert, P.W. 1961. Progress Report to Office of Naval Research, January 1, 1960-December 31, 1961, Arlington, Va.

_____, ed. 1963a. *Sharks and Survival.* D.C. Heath and Co., Boston.

_____. 1963b. The AIBS Shark Research Panel. Pages 505-507 in P.W. Gilbert, ed. *Sharks and Survival.* D.C. Heath and Co., Boston.

_____. 1963c. Advice to those who frequent, or find themselves in, shark-infested waters. Pages 501-503 in

P.W. Gilbert, ed. Sharks and Survival. D.C. Heath and Co., Boston.

_____. 1970. Studies on the Anatomy, Physiology, and Behavior of Sharks. Unpublished final report to Office of Naval Research (code 484), Arlington, Va.

_____. 1976. An evaluation of some chemical, biological, and physical agents tested for their effectiveness as shark deterrents. Page 19 in W. Seaman, Jr., ed. Sharks and Man: A Perspective. Florida Sea Grant Program Report 10, Gainesville, Fla.

Gilbert, P.W., and C.K. Gilbert. 1973. Sharks and shark deterrents. Underwater J. 5(2):69-79.

Gilbert, P.W., and S. Springer. 1963. Testing shark repellents. Pages 477-494 in P.W. Gilbert, ed. Sharks and Survival. D.C. Heath and Co., Boston.

Goodman, L.S., and A. Gilman. 1970. The Pharmacological Basis of Therapeutics. 4th ed. Macmillan, New York.

Gruber, S.H. 1981. Shark repellents: perspectives for the future. Oceanus 24(4):72-76.

Gruber, S.H., and E. Zlotkin. 1982. Bioassay of surfactants as shark repellents. Nav. Res. Rev. 34(2):18-27.

Guyton, A.C. 1976. Textbook of Medical Physiology. W.B. Saunders, Philadelphia.

Hall, H. 1981. Anatomy of shark bite. Skin Diver Mag. 30 (4):18-19.

Halstead, B.W. 1959. Dangerous Marine Animals. Cornell Maritime Press, Maryland.

_____. 1978. Poisonous and Venomous Marine Animals of the World, rev. ed. Darwin Press, Princeton, N.J.

Hara, T.J. 1971. Chemoreception. Pages 79-120 in W.S. Hoar and D.J. Randall, eds. Fish Physiology, Vol. 5. Academic Press, New York.

Harding, J.H. 1975. Filming the whaler shark. Sea Frontiers 21(2):76-81.

Hasler, A.D. 1957. Olfactory and gustatory senses of fishes. Pages 187-209 in M.E. Brown, ed. The Physiology of Fishes, Vol. 2, Behavior. Academic Press, New York.

Hasler, A.D., and W.J. Wisby. 1959. Method of Repelling Fish Comprising Treating with Potassium Phenyl Acetate. U.S. Patent Office. Pat. 2,880,133, March 31, 1959.

Hasler, A.D., A.T. Scholz, and R.M. Horrall. 1978. Olfactory imprinting and homing in salmon. Am. Sci. 66:347-355.

Hass, H. 1966. Ich Fotografierte in den 7 Meeren. Im Herring-Verlag, Seebruck am Chiemsee, Germany.

Hiatt, R.W., J.J. Naughton, and D.C. Matthews. 1953. Effects of chemicals on a schooling fish, Kuhlia sandvicensis. Biol. Bull. 104:28-44.

_____. 1957. Methods of Dispersing Schools of Fish. Final Progress Report, Office of Naval Research (Biology Branch), Arlington, Va.

Hobson, E.S. 1963. Feeding behavior in three species of sharks. Pac. Sci. 17(2):171-194.

Hobson, E.S., F. Mautin, and E.S. Reese. 1961. Two shark incidents at Eniwetok Atoll, Marshall Islands. Pac. Sci. 15 (4):605-609.

Hodgson, E.S., and R.F. Mathewson. 1978. Electrophysiological studies of chemoreception in elasmobranchs. Pages 227-267 in E.S. Hodgson and R.F. Mathewson, eds. Sensory Biology of Sharks, Skates, and Rays. Office of Naval Research, Arlington, Va.

Idler, D.R., U.H.M. Fagerlund, and H. Mayoh. 1956. Olfactory perception in migrating salmon. 1. L-serine, a salmon repellent in mammalian skin. J. Gen. Physiol. 39(6):889-892.

Jernejcic, F. 1969. Use of emetics to collect stomach contents of walleye and largemouth bass. Trans. Am. Fish. Soc. 98(4):698-702.

Johnsen, P.B., and A.D. Hasler. 1980. The use of chemical cues in the upstream migration of coho salmon, Oncorhynchus kisutch Walbaum. J. Fish. Biol. 17:67-73.

Johnson, C.S. 1972. Countermeasures to shark attack. Pages

123-141 in G.V. Pickwell and W.E. Evans, eds. Handbook of Dangerous Animals for Field Personnel. Naval Undersea Center, San Diego, TP 324.

_____. 1976. Anti-shark devices and testing methods at Naval Undersea Center. Page 17 in W. Seaman, Jr., ed. Sharks and Man: A Perspective. Florida Sea Grant Program Report 10, Gainesville, Fla.

Johnson, C.S., and J.F. Fish. 1974. Letter-to-editor. Skin Diver Mag. (August):7.

Johnson, R.H. 1978. Sharks of Polynesia. Editions du Pacifique, Papeete, Tahiti.

Johnson, R.H., and D.R. Nelson. 1973. Agonistic display in the gray reef shark, Carcharhinus menisorrah, and its relationship to attacks on man. Copeia 1973:76-84.

Katsuki, Y., K. Yanagisawa, A.L. Tester, and J. Kendall. 1969. Shark pit organs: responses to chemicals. Science 163:405-407.

Kinnel, R.B., R.K. Dieter, J. Meinwald, D. Van Engen, J. Clardy, T. Eisner, M.O. Stallard, and W. Fenical. 1979. Brasilenyne and cis-dihydrorhodophytin: antifeedant medium-ring haloethers from a sea hare (Aplysia brasiliana). Proc. Natl. Acad. Sci. USA 76(8): 3576-3579.

Kleerekoper, H. 1969. Olfaction in Fishes. Indiana University Press, Bloomington.

_____. 1978. Chemoreception and its interaction with flow and light perception in the locomotion and orientation of some elasmobranchs. Pages 269-329 in E.S. Hodgson and R.F. Mathewson, eds. Sensory Biology of Sharks, Skates, and Rays. Office of Naval Research, Arlington, Va.

Klimley, A.P. 1974. An inquiry into the causes of shark attacks. Sea Frontiers 20(2):66-75.

Klimley, A.P., and A.A. Myrberg, Jr. 1979. Acoustic stimuli underlying withdrawal from a sound source by adult lemon sharks, Negaprion brevirostris (Poey). Bull. Mar. Sci. 29(4):447-458.

Langguth, A.F. 1972. Gas injection shark dart. Am. Soc. Mech. Eng. Pub. 72-WA/OCT. -10.

Llano, G.A. 1963. Open-ocean shark attacks. Chapter 13, pp. 369-386 in P.W. Gilbert, ed. Sharks and Survival. D.C. Heath and Co., Boston.

Mathewson, R.F., and E.S. Hodgson. 1972. Klinotaxis and rheotaxis in orientation of sharks toward chemical stimuli. Comp. Biochem. Physiol. 42(1A):79-84.

McCosker, J.E. 1981. Great white shark. Science 81 2(6):42-51.

McKenney, J. 1975. Sea-way fast-load shark shooter. Skin Diver Mag. 24(4):46-47, 62-63.

McKibben, J.N., and D.R. Nelson. 1981. A portable, real-time, X-Y plotting system for ultrasonic tracking of fish. Pages 105-115 in F.M. Long, ed. Proceedings of Third International Conference on Wildlife Biotelemetry. ICWB, University of Wyoming, Laramie.

McNair, R. 1975. Sharks I have known. Skin Diver Mag. 24(1):52-57.

Miller, D.J., and R.S. Collier. 1981. Shark attacks in California and Oregon, 1926-1979. Calif. Fish Game. 67(2): 76-104.

Moss, S.A. 1977. Feeding mechanisms in sharks. Am. Zool. 17(2):355-364.

_____. 1981. Shark feeding mechanisms. Oceanus 24(4):23-29.

Mrosovsky, N., and D.F. Sherry. 1980. Animal anorexias. Science 207:837-842.

Myers, C.W., and J.W. Daly. 1976. Preliminary evaluation of skin toxins and vocalizations in taxonomic and evolutionary studies of poison-dart frogs (Dendrobatidae). Bull. Am. Mus. Nat. Hist. 157(art. 3):173-262.

Myers, C.W., J.W. Daly, and B. Malkin. 1978. A dangerously toxic new frog (Phyllobates) used by Emberia Indians of western Columbia, with discussion of blowgun fabrication and dart poisoning. Bull. Am. Mus. Nat. Hist. 161(art. 2):307-366.

Myrberg, A.A., Jr. 1978. Underwater sound--its effect on the behavior of sharks. Pages 391-417 in E.D. Hodgson and R.F. Mathewson, eds. Sensory Biology of Sharks,

Skates, and Rays. Office of Naval Research, Arlington, Va.

Myrberg, A.A., Jr., and S.H. Gruber. 1974. The behavior of the bonnethead shark, Sphyrna tiburo. Copeia 1974: 358-374.

Myrberg, A.A., Jr., C.R. Gordon, and A.P. Klimley. 1978. Rapid withdrawal from a sound source by open-ocean sharks. J. Acoust. Soc. Am. 64(5):1289-1297.

Nelson, D.R. 1969a. The silent savages. Oceanus 1(4):8-22.

_____. 1969b. Meeting the shark underwater. Underwater Nat. 6(2):13-26.

_____. 1981. Aggression in sharks: is the gray reef shark different? Oceanus 24(4):45-55.

Nelson, D.R., and R.H. Johnson. 1972. Acoustic attraction of Pacific reef sharks: effect of pulse intermittency and variability. Comp. Biochem. Physiol. 42(1A):85-96.

_____. 1980. Behavior of the reef sharks of Rangiroa, French Polynesia. Natl. Geogr. Soc. Res. Rep. 12:479-499.

Parker, D.O. 1971. Effect of molecular structure on the toxicity of some nonionic surfactants to the goldfish, Carassius auratus (Cypriniformes: Cyprinidae). Masters thesis, California State University, Long Beach. 92 pp.

Primor, N., J. Parness, and E. Zlotkin. 1978. Pardaxin: the toxic factor from the skin secretion of the flatfish Pardachirus marmoratus (Soleidae). Pages 539-547 in P. Rosenberg, ed. Toxins: Animal, Plant, and Microbial. Pergamon Press, Oxford.

Randall, J.E. 1963. Dangerous sharks of the western Atlantic. Pages 339-361 in P.W. Gilbert, ed. Sharks and Survival. D.C. Heath and Co., Boston.

Randall, J.E., and G.S. Helfman. 1973. Attacks on humans by the blacktip reef shark (Carcharhinus melanopterus). Pac. Sci. 27:226-238.

Scholz, A.T., R.M. Horrall, J.C. Cooper, and A.D. Hasler.

1976. Imprinting to chemical cues: the basis of home-stream selection in salmon. Science 192:1247-1249.

Schultz, L.P. 1963. Attacks by sharks as related to activities of man. Pages 425-452 in P.W. Gilbert, ed. Sharks and Survival. D.C. Heath and Co., Boston.

Sheldon, R.E. 1909. The reactions of dogfish to chemical stimuli. J. Comp. Neurol. 19:273-311.

Shorey, H.H. 1977. Pheromones. Pages 137-163 in T.A. Sebok, ed. How Animals Communicate. Indiana University Press, Bloomington.

Sobotka, H. 1962. The Effect of Holothurin from Sea Cucumbers as a Shark Repellent. Unpublished progress report.

_____. 1965. Comparative biochemistry of marine animals. BioScience 15:583-584.

Springer, S. 1967. Social organization of shark populations. Pages 149-174 in P.W. Gilbert, R.F. Mathewson, and D.P. Rall, eds. Sharks, Skates and Rays. Johns Hopkins Press, Baltimore.

Starck, W.A. 1974. Letter-to-editor. Skin Diver Mag. (October):6.

Starck, W.A. and A. Anderson. 1979. The Blue Reef--A Report from Beneath the Sea. Alfred A. Knopf, New York.

Steinberg, M.A. 1960. Chemical Analysis of Shark-Repelling Substances. Progress Report, Office of Naval Research, Arlington, Va.

Taylor, V. 1981a. Jawbreaker for sharks. Natl. Geogr. 159 (5):664-667.

_____. 1981b. Anti-shark armor that works. Skin Diver Mag. 30(10):23-27.

_____. 1981c. Shark suit meets Jaws. Skin Diver Mag. 30 (11):19-20.

Teeter, J. 1980. Pheromone communication in sea lampreys (Petromyzon marinus): implications for population management. Can. J. Fish. Aquat. Sci. 37(11):2123-2132.

Tester, A.L. 1962. A Summary of Research on Sharks. Unpublished report. Government-Industry Tuna Meeting, La Jolla, Calif., January 9-10.

_____. 1963. Olfaction, gustation, and the common chemical sense in sharks. Pages 255-282 in P.W. Gilbert, ed. Sharks and Survival. D.C. Heath and Co., Boston.

Tester, A.L., and G.J. Nelson. 1967. Free neuromasts (pit organs) in sharks. Pages 503-531 in P.W. Gilbert, R.F. Mathewson, and D.P. Rall, eds. Sharks, Skates, and Rays. Johns Hopkins Press, Baltimore.

Thompson, T.E. 1960. Defensive adaptions in opisthobranchs. J. Mar. Biol. Assoc. U.K. 39:123-134.

Thompson, T.E., and D.J. Slinn. 1959. On the biology of opisthobranch Pleurobranchus membraneceus. J. Mar. Biol. Assoc. U.K. 38:507-524.

Tuve, R.L. 1963. Development of the U.S. Navy "Shark Chaser" chemical shark repellent. Chapter 17, pages 455-464 in P.W. Gilbert, ed. Sharks and Survival. D.C. Heath and Co., Boston.

Uexkull, J. von. 1895. Vergleichend-sinnesphysiologische Untersuchungen I. Uber die Nahrungsaufnahme des Katzenhais. Z. Biol. 32:548-566.

Ulrich, R.E. 1966. Pain as a cause of aggression. Am. Zool. 6:643-662.

Wallett, T. 1978. Shark Attack and Treatment of Victims in Southern African Waters. Prunell and Sons, Capetown.

Waterman, S. 1979. Focus on sharks. Skin Diver Mag. 28 (3):8-13.

Whitley, G.P., and G.H. Payne. 1947. Testing a shark repellent. Aust. Zool. 2(2):151-157.

Zahuranec, B.J., ed. 1975. Shark Research: Present Status and Future Direction. Office of Naval Research, ONR Report ACR-208, Arlington, Va.

William O. McClure, Joseph V. Martin

3. On the Probability of Finding Shark Repellents in Marine Organisms

The sea has provided us with a rich store of potentially useful bioactive substances. With several colleagues, we have sought extracts of marine organisms for substances possessing specific properties. For example, Lowell Hager's group has found halogenated natural products (Hager et al., 1974), as well as the enzymes that catalyze the oxidation of halides and their subsequent biosynthetic incorporation. Kenneth Rinehart's group has obtained compounds active against yeast, fungi, bacteria, viruses, and cells (Rinehart et al., 1980); some of the lattermost may be effective as anti-cancer agents.

Our group has sought neurotoxins that specifically affect the release of neurotransmitter from the presynaptic terminal, with the goal of using the toxins as tools in our basic study of the release process (Martin and McClure, 1979, 1980). Because an effective shark repellent may inhibit nervous system function, our research may be quite relevant to the detection and eventual isolation of anti-shark agents. In this report we briefly describe our methods and success, paying particular attention to points of philosophy that may be useful in the search for a shark repellent.

Methods

We collected sample organisms from intertidal or subtidal regions, usually by using SCUBA. We took organisms present in large numbers, because we anticipated requiring kilogram amounts of those that proved active. Whenever possible, we tried to minimize ecological damage to an area by taking small initial samples (2-5 g) for preliminary assessment of activity. Upon detection of activity, we returned for a large sample.

Although we could not always follow this collection

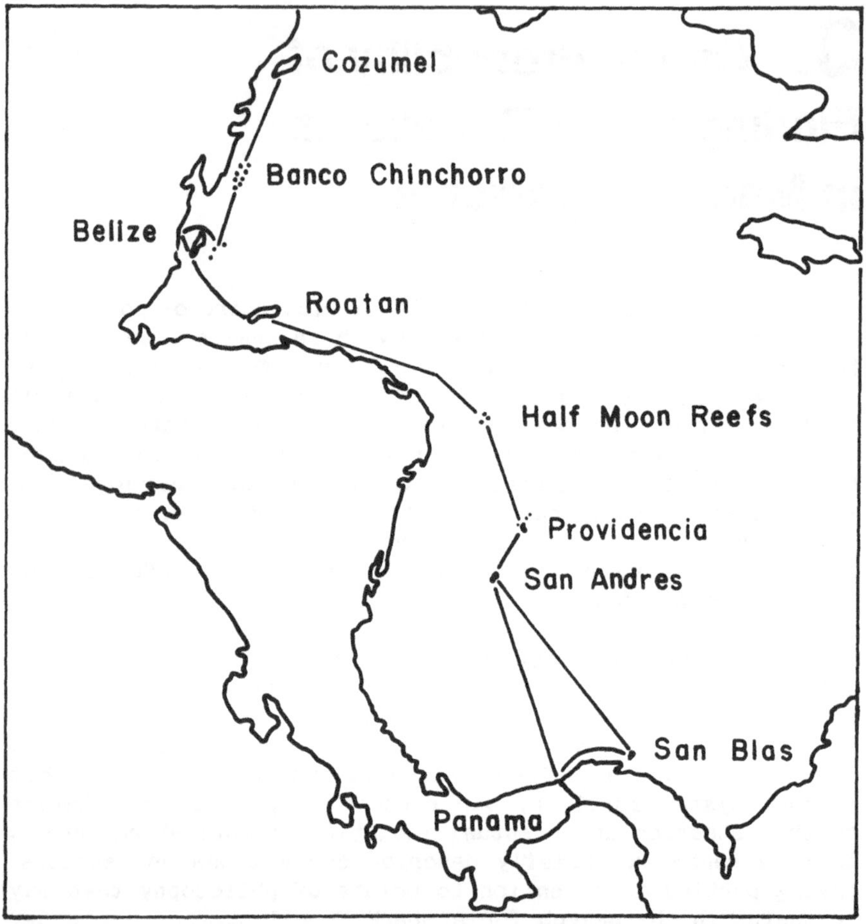

Figure 1. Map of cruise track of the research vessel Alpha Helix on the Caribbean expedition of 1978. The vessel embarked from Panama, traveled to San Andres, and then proceeded north to Cozumel. On the return trip to Panama, the same track was followed except for a diversion to the San Blas islands. Collections were obtained at 13 sites (not shown; see Rinehart et al., 1980).

procedure, we did use it extensively during an expedition on the research vessel Alpha Helix in 1978. In the first phase of the trip, we moved north from the Panama Canal to Cozumel (see Figure 1), making numerous collections of small samples and analyzing them for activity. In the second phase, we retraced our path and made larger collections of the active organisms.

Tissues were cleaned of detritus, after which a representative sample of about 2 g was weighed and homogenized for 30 sec in a Virtis Model 23 rotating-blade homogenizer in 10mL of an ice-cold extraction medium that mimicked the composition of fish serum. The concentration of salts in the medium was: NaCl (160mM), KCl (6.4mM), Na acetate (34mM), $MgCl_2$ (6.0mM), $CaCl_2$ (5.4mM), glucose (10mM), and NaHEPES (10mM), at a pH of 7.4. The sample homogenizations were highly variable, but most provided a creamy near-homogenous extract. Some organisms, such as hard corals, were homogenized with previously dulled blades. In such cases (animals having a substantial amount of hard tissue), the biological activities detected later were due to an amount of soft tissue considerably less than the usual 2 g.

After homogenization, the samples were sedimented for 5 min in a Sorvall RC2-B centrifuge at 35,000 xg. The supernatants were removed and frozen prior to assay. If a lipid layer was present above the aqueous sample, we used a Pasteur pipette to remove the aqueous phase with as little contamination as possible.

In cases that required alteration of the ionic composition of the samples, we used "minicolumns" of Sephadex G-25 packed with a bed volume of 5 mL in the barrel of a disposable hypodermic syringe. With plasticware of the proper configuration, we could elute the columns with buffer by centrifuging gently in a clinical centifuge (Paulson, 1973). As usual with Sephadex, the larger molecules emerge from the column in an aliquot of the buffer in which the column was originally equilibrated, thereby effectively transfering the larger molecules from the original buffer into the column buffer. Recovery of proteins examined as standards usually exceeded 90%, and several samples could be processed in parallel, which made the method attractive for screening purposes.

We monitored release of neurotransmitter by using synaptosomes from fish freshly caught by hand-spearing. A typical catch brought 4 to 8 small fish to the shipboard laboratory within 30 min of death. Of several species examined, the reef squirrelfish (holocentridae myriprystis jacobus) seemed the most suitable because it was easily available and provided high-quality synaptosomes. Although we did not study the point exhaustively, we did notice a strong qualitative relationship between the size of the eye of the fish and the quality of the synaptosomes. Small-eyed fish had small brains, even if large bodied (e.g., triggerfish, angelfish), and produced only a small amount of synaptosomes of poor quality. Large-eyed fish, however, had relatively large brains and produced synaptosomes nearly equal in quality to those from the usual laboratory rat.

We prepared the synaptosomes by a modification of the method of Whitaker and Greengard (1971) and immediately labeled them by incubation for 30 min at 25° with 10 μCi of [methyl-^{3}H]choline. Then we separated them into several aliquots, each of which would suffice for one batch of assays, and placed them on ice (Yoshino et al., 1980). To carry out a set of release assays, we thawed a number of frozen extracts (usually 10 to accommodate our filtration apparatus) and clarified them by centrifugation at 35,000 xg for 5 min. An aliquot of synaptosomes was warmed at 25°C for 3 min, and a volume of this suspension corresponding to 20 mg of the initial wet weight of brain was applied to a 24 mm Millipore AAWP filter. Gentle vacuum was applied to pull the suspending liquid through the filter, after which the filters were washed with 10 sequential aliquots of 2.5 mL each of extraction medium. To examine the action of an extract, a volume of 1.0 mL was then applied to the washed synaptosomal bed on the filter and allowed to remain there for 1 min before being drawn through and collected. The filters were washed with a final 1.5 mL of extraction medium, which was drawn through 1 min later and combined with the immediately preceding sample to form the experimental sample.

In each set of 12 samples only 10 extracts were tested. The 2 remaining ports were used as controls in which extraction medium was substituted for the extract. Results of each sample were expressed in terms of the average value of the controls. We then checked for proper functioning of the synaptosomes by substituting as a sample either extraction medium containing 67 mM K^+ ions or an extract previously found to possess releasing activity. If high levels of K^+ were used, isosmotic conditions were maintained by reducing the concentration of NaCl. We determined radioactivity in the experimental samples through liquid scintillation spectrometry by using a Beckman model 7500 spectrometer (at the University of Southern California) or a Beckman model LS-100 spectrometer (on board ship) and a scintillant of Triton and xylene (Anderson and McClure, 1973). We calculated lactate dehydrogenase activity in experimental samples by following the decreases in absorbance at 340 nm produced by the oxidation of NADH by pyruvate, according to the method of Goldberg (1975).

In many experiments we decided to fractionate the released radioactivity to compare the amounts of acetylcholine and choline. We tried several separation methods, each of which produced similar results. The procedure we now use involves conversion of choline to phosphorylcholine using choline kinase and ATP, following the method of Goldberg and McCaman (1973); then acetylcholine and the newly formed phosphoryl choline can easily be separated by their different charges.

In most cases we used the method of Lowry et al. (1951) to make protein determinations. When greater sensitivity was required, we used the dye-binding assays of Bradford (1976). All reagents were reagent grade unless otherwise specified. Distilled water was used throughout.

Results

During the 1978 Caribbean cruise of the Alpha Helix, we analyzed extracts from 313 animal samples for their ability to stimulate the release of radioactivity from fish brain synaptosomes that had been previously labeled with [^{3}H]choline. These extracts represent about 240 species, although the exact number is not known because taxonomic identifications are still incomplete. Virtually all the extracts caused some release of radioactivity (see Figure 2). A distribution profile of releasing activity indicates a mean release of 327% of control; that is, the average sample released 3.27 times as much radioactivity as did a control of the extraction medium alone. Such a high average level of releasing activity in animals suggests that it is not of importance in the search for a specific neurotoxin.

Of greater significance to us are the samples that exhibit releasing activity well in excess of the mean. Of the organisms cited in Figure 2, 39% gave release exceeding 300% of control, and 14% gave release exceeding 550% of control. The phylogenetic groups to which these active samples belong are relatively few, and include primarily six major taxa (see Table 1). The sea cucumbers (holothurians) and the anemones showed an especially high incidence of active species. By far the greatest number of active species was found in the sponges (Porifera), which reflects the large number of sponges in the Caribbean.

We also examined plants for releasing activity. A

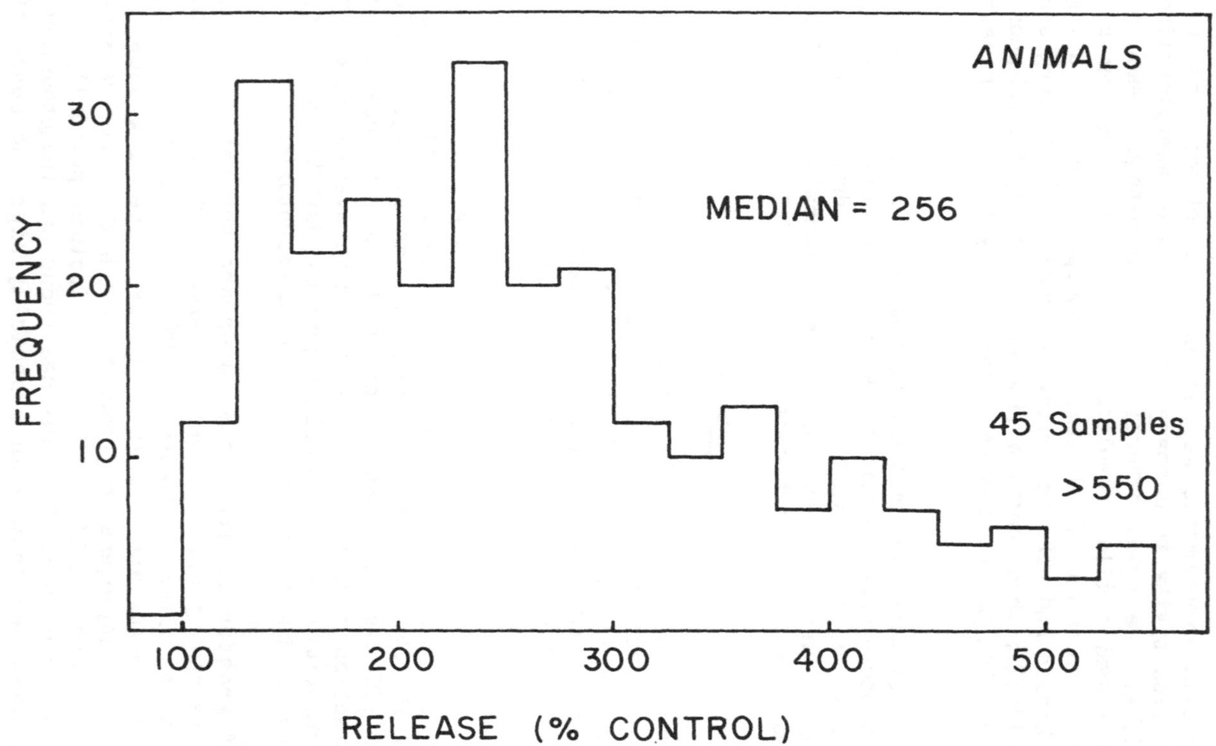

Figure 2. Distribution profile of releasing activity for extracts of the animals collected on the Alpha Helix Caribbean expedition of 1978. Release is defined as percentage of control; that is, a release of 200 indicates a sample that released 2.0 times as much radioactivity as did a toxin-free control.

Table 1. Phylogenetic distribution of releasing activity in extracts of animals sampled on the 1978 Alpha Helix Caribbean expedition.[a]

Animals	Number of Tested Species	Number of Species[b] Active	Percentage of Species Active
Porifera	131	37	28
Gorgonians	32	11	34
Hard corals	26	11	42
Anemones	7	4	57
Sea urchins	6	2	33
Holothurians	7	6	86
TOTAL	209	71	34

a. Only taxa represented by significant numbers of samples or significant incidence of activity are presented.

b. "Active" = possessing release of greater than 300% of control.

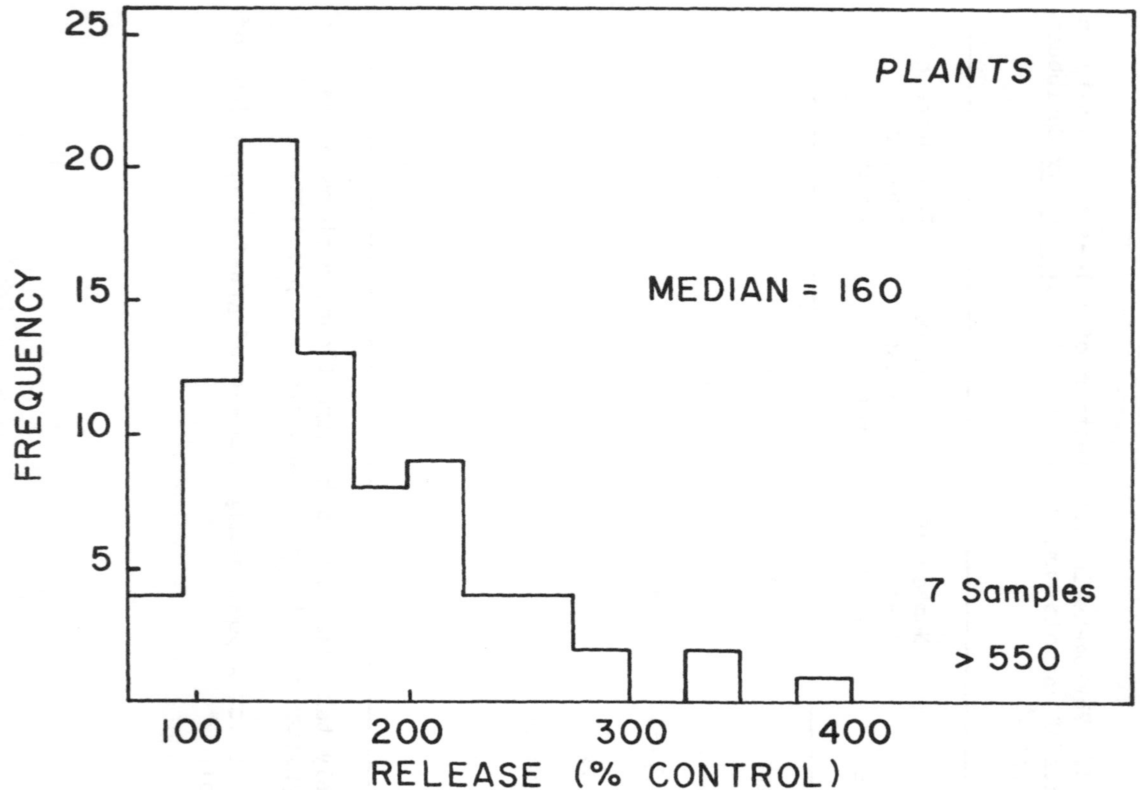

Figure 3. Distribution profile of releasing activity for extracts of the plants collected on the Alpha Helix Caribbean expedition of 1978. See caption to Figure 2 and the text for details.

distribution profile for 87 samples, which represent 80 species, indicated that most plants do not significantly stimulate release (see Figure 3). The average release was 212% of control, much less than the 327% seen with animals. Marine plants do contain active releasing agents, however. In this set of samples, 8% yielded release values greater than 550%. Among the phylogenetic groups with activity, the common algal classes are represented about equally (Table 2).

We briefly examined a number of aspects of the release stimulated by these active extracts in an attempt to eliminate from consideration those extracts not stimulating release in a physiologically relevant manner. In the first winnowing of the number of active samples, we examined the level of lactate dehydrogenase (LDH) in the experimental sample. If an extract causes simultaneous loss of radioactivity and LDH--an enzyme commonly used as a marker for the cytosol--the extract probably causes release simply by destroying the integrity of the cell membrane. An extract with these properties can safely be excluded from further consideration. Of the extracts that caused release in excess of 550% of control, about one-third also caused release of significant amounts of LDH from synaptosomes (see Table 3). The remaining samples became subjects of further study.

Experience with a few previous marine extracts, with a toxin from a terrestrial beetle (McClure et al., 1980), and with the releasing activity in black widow spiders, suggests that many of the toxins of interest to us will be proteins. To examine this point as well as remove inorganic ions that may stimulate release, we passed aliquots of active extracts over columns packed with Sephadex, a molecular wieve. This procedure transfers the high molecular-weight compounds (less than 5000 daltons) into the buffer with which the Sephadex was equilibrated--either extraction medium aboard the Alpha Helix or a rat Ringer's solution at the university laboratory. Separation of the small molecules from the larger ones indicated that small molecules were responsible for the observed activity in about one-third of all extracts. The samples in which activity resides in small molecules may contain either an inorganic ion or an organic toxin. L.V. Dunkerton and B. McKillican (personal communication, 1981) have fractionated some of the active samples of smaller molecular weight, and have partially purified small toxins from them.

We have not further considered the material of small molecular weight, preferring instead to study the protein-containing fractions. We have therefore included in our screening only those samples with activity in the components

Table 2. Phylogenetic distribution of releasing activity in extracts of algae sampled on the 1978 Alpha Helix Caribbean expedition.[a]

Algae	Number of Tested Species	Number of Species[b] Active	Percentage of Species Active
Red	32	13	41
Brown	15	5	33
Green	28	7	25
Blue-green	2	1	50
TOTAL	77	26	34

a. Only taxa represented by significant numbers of samples or significant incidence of activity are presented.

b. "Active" = possessing release of less than 200% of control.

Table 3. Collected properties of selected extracts of animals sampled on 1978 Alpha Helix Caribbean expedition.

Sample Number[a]	Release of Radio-activity	Release of LDH[b]	Acetylcholine in Released Material (% of Total Radioactivity)[c]	Comments
323	-0.8	3.6	35	inactive
126	4.0	5.5	19	inactive
141	9.9	-0.6	16	lowactivity; choline release
254	6.0	6.5	36	activity probably lytic
250	13.7	22.6	45	low activity; probably lytic
178	68.3	68.9	58	probably lytic
183	33.2	12.2	59	releaser
113	65.3	28.3	56	good releaser
51	82.9	22.3	59	good releaser
335	100	23.4	49	good releaser; low acetylcholine
Veratridine	17.5		71	known releaser
67mM K^+	13.0	7.9	67	known releaser
Control	0	0	50	

a. Only desalted samples are considered. The samples are selected to provide examples of different cases presented in the text.

b. To allow comparison, results are expressed as percent of the change between control and maximal release, by the equation:

$$\text{release } (\%) = \left(\frac{\text{observed release} - \text{control release}}{\text{maximal release} - \text{control release}}\right) \times 100$$

Maximal release was determined from lysates of the bed of synaptosomes. Control values, expressed as percent of the maximal release, were 6-7% for radioactivity and 13-14% for LDH. The standard deviation for each value is +/- 5-8%.

c. Expressed without correction for control. The standard deviation is +/- 3%.

of high molecular weight. However, a procedure directed toward finding shark repellents would probably not want to consider proteins, for these agents would be difficult to obtain in commercially useful quantities, no matter how useful they are as tools for laboratory research. The search for shark repellents would probably find the toxic activities of small-molecular-weight compounds more useful.

Release that we have so far considered has been couched in terms of total radioactivity. Since both acetylcholine and choline are labeled in the synaptosomes and are released simultaneously under physiological conditions, we tried to define the nature of the radioactive compounds released by the remaining active extracts. We used choline kinase and ATP to convert choline to phosphorylcholine, which is easily separated from acetylcholine. We also precipitated acetylcholine and choline (Stavinoha and Weintraub, 1974) as their periodide adducts, which can be solubilized, converted to their chlorides, and separated by chromatography.

Data obtained by both methods are in good qualitative agreement, although we encountered several artifacts which rendered the periodide method of lesser value. Several of the extracts have interesting properties (see Table 3). For purposes of comparison, release stimulated by 67 mM K^+ ions yields a filtrate in which 65-70% of the radioactivity is present as acetylcholine. Samples only slightly richer in acetylcholine (75-80%) are observed when 1 x 10^{-5} M veratridine is used to stimulate release. Some extracts can release material rich in choline (see Table 3). The data indicate that the release of choline may occur by pathways different from those by which acetylcholine is released, despite the fact that both substances are released simultaneously and different mechanisms may be differentially stimulated by the extracts.

Discussion

Our search in extracts of marine organisms for neurotoxins with one specific property, that of releasing acetylcholine from the presynaptic terminal, has been successful. Although much work remains to be done with these extracts, the presence of high levels of releasing activity is clearly established. The success of this project makes it likely that extracts of marine organisms can yield activities directed toward other phenomena. For example, a very similar approach has recently led to the discovery of several organisms that possess very active antiviral compounds, and to the isolation and determination of the structures of a new family of anti-viral agents, the didemnins (Rinehart et al.,

1981). We have subsequently been associated, through studies conducted by persons in the laboratories of either Lowell Hager, Kenneth Rinehart, or at the University of Southern California, with successful searches in marine extracts for 12 different activities; thus, similar investigations for others would be appropriate.

With respect to the specific goal of finding a shark repellent by this means, several points must be kept in mind. First, a rapid and meaningful assay is an absolute requirement. If the assay is too slow, it will take too long to find the desired extracts. Although sheer physical labor can overcome this problem in some cases, there are some activities for which the duration of each assay becomes a significant limiting factor. Further, the assay must be relevant; that is, it must measure a property directly related to that which is desired.

In some cases the assay may be a compromise among several factors, such as expense, time, selectivity, sensitivity, etc. The experiments described in this paper are an example of a compromise in the assay. To find toxins unequivocally active at the presynaptic terminal, we should use a more direct assay than release from synaptosomes; that is, we should use an intact presynaptic terminal. The "best" assay may involve the use of intra-cellular microelectrodes to monitor the frequency of miniature endplate potentials, which are usually taken as the best single index of presynaptic function. One could, at least in theory, apply several extracts sequentially to a properly functioning nerve-muscle preparation and observe the frequency of miniature endplate potentials after each application. In practice, however, this procedure is excessively tedious, and thus not a practical screening assay. As a result, we have employed the synaptosomal assay, recognizing that it is a compromise between ease of handling and relevance. Once active extracts are detected, we must examine their activity at the neuromuscular junction, which we are now doing.

In searching for a shark repellent, we are optimistic that assays carried out at less than the level of the whole animal will be particularly useful. Certainly, toxicity to small fish in laboratory aquariums is a necessary compromise between relevance and ease, but the fish toxicity assay described by Gerald Bakus (see chapter 10) is a good beginning and deserves support. Use of this assay, coupled with definitive tests on the most active extracts, could bring detection and isolation of molecules with activity as shark repellents.

Certain practical aspects of the desired molecule should also be borne in mind. In the case of a shark repellent, at least three constraints must be considered: First, the molecule should be active at very low concentration, for it will probably have to be used by dispersal into a reasonably large volume of seawater. Second, the molecule must be quite stable, for packages of repellent are likely to be stored for long periods of time (years) on ships and water-based aircraft. A similar constraint exists with respect to the formulation in which the repellent will be given to users. Finally, the best repellents will probably be of low molecular weight, for two reasons: (a) an effective repellent will be required in large amounts, which can probably be more easily obtained by chemical synthesis than by isolation; and (b) the initial screening program may uncover molecules possessing characteristics that are very good, but not quite good enough. If the structure of such a "near-miss" repellent is known, chemical derivatives can be made. The new molecules may have different properties, and may be superior to their progenitors. The pharmaceutical industry has long used the approach of trial-and-error modification of molecules already known to possess some activity. The same approach should be useful with shark repellents. Since small molecules can usually be modified more easily than large ones, a low-molecular-weight toxin would be preferable.

Research directed toward the detection and isolation of shark repellents is now being pursued in the Marine Bioactive Substances Program at the University of Southern California. By using the concepts developed in both the Bakus paper (see chapter 10) and this paper, we should be able to forge useful tools for mankind from the molecules used for eons as nature's weapons.

Acknowledgments

It is a pleasure to thank K.L. Rinehart, Jr., and L.P. Hager for permission to quote from their work in the symposium address on which this manuscript is based. We wish also to thank the National Science Foundation for support, as well as the NIEHS (ES 001965-03) and Nelson Research and Development, Irvine, California. Preliminary communications of portions of this work have been published (Martin and McClure, 1979, 1980).

References Cited

Anderson, L.E., and W.O. McClure. 1973. An improved scintillation cocktail of high solubilizing power. Anal. Biochem. 51: 173-179.

Bradford, M. 1976. A rapid and sensitive method for the quantitation of microgram quantities of protein utilizing the principle of protein-dye binding. Anal. Biochem. 72:248-254.

Goldberg, A.L., and R.E. McCaman. 1973. The determination of picomole amounts of acetylcholine from mammalian brain. J. Neurochem. 20:1-8.

Goldberg, E. 1975. Lactate dehydrogenase-X from mouse testes and spermatozoa. Pages 318-323 in W.P. Wood, ed. Methods in Enzymology, Vol. XLI. Academic Press, New York.

Hager, L.P., R.H. White, P.F. Hollenberg, D.L. Doubek, R.C. Brusca, and R. Guerrero. 1974. A survey of organic halogens in marine organisms. Pages 421-425 in H.H. Webber and D. Ruggieri, eds. Food Drugs From the Sea. Marine Technology Society, Washington, D.C.

Lowry, O.H., H.J. Rosebrough, A.L. Farr, and R.J. Randall. 1951. Protein measurement with the Folin-phenol reagent. J. Biol. Chem. 183:265-275.

Martin, J.C., and W.O. McClure. Presynaptic neurotoxins from the sea. 1979. Trans. Am. Soc. Neurochem. 10:99.

_____. 1980. Analyses of marine presynaptic neurotoxins. Trans. Am. Soc. Neurochem. 11:85.

McClure, W.O., B.C. Abbott, D.E. Baxter, T.H. Hsiao, L.S. Satin, A. Siger, and J.E. Yoshino. 1980. Leptino tarsin: a presynaptic neurotoxin that stimulates release of acetylcholine. Proc. Natl. Acad. Sci. USA 77: 1219-1223.

Paulson, J.C. 1979. Detection and quantitation of drug-induced inhibition of rapid axoplasmic transport: microtubules and axoplasmic transport. Ph.D. Thesis, University of Illinois.

Rinehart, K.L., Jr., P.D. Shaw, L.S. Shield, J.B. Gloer, G.C. Harbour, M.E.S. Koker, D. Samian, R.E. Schwartz, A.A. Tymiak, D.L. Weller, G.T. Carter, M.H.G. Munro, R.G. Hughes, Jr., H.E. Renis, E.G. Swynenberg, D.A. Stringfellow, J.J. Vavra, J.H. Coats, G.E. Zurenko, S.L. Kuentzel, L.H. Li, G.J. Bakus, R.C. Brusca, L.L. Craft, D.N. Young, and J.L. Connor. 1980. Marine natural products as sources of antiviral, antimicrobial, and antineoplastic agents. Pure Appli. Chem. 53:795-817.

Stavinoha, W.B., and S.T. Weintraub. 1974. Estimation of choline and acetylcholine by pyrolysis gas chromatography. Anal. Chem. 46: 751-760.

Whittaker, V.P., and P. Greengard. 1971. The isolation of synaptosomes from the brain of a teleost fish, *Centriopristes striatus.* J. Neurochem. 18: 173-176.

Yoshino, J.E., D.E. Baxter, T.H. Hsiao, and W.O. McClure. 1980. Release of acetylcholine from rat brain synaptosomes stimulated with leptinotarsin, a new neurotoxin. J. Neurochem. 38:635-642.

Samuel H. Gruber

4. Shark Repellents: Protocols for a Behavioral Bioassay

Attacks by sharks on humans involve a complicated series of behavioral steps starting with detection by one or more of the shark's sensory systems and continuing through orientation, approach, and finally actual contact with the victim. To intervene and successfully block an attack, we must affect the shark's detection or orientation processes so as to alter radically its motivation.

Clearly, altering sensory input is one of the most direct and rapid pathways to interrupt shark attack. Therefore, the testing and specification of aversive or repellent stimuli designed to interrupt sensory pathways fall directly within the domain of sensory psychophysics and comparative psychology--disciplines organized, in large part, to elucidate the relationship between stimulus fields and integrated responses. For example, behavioral toxicology, which employs the learning paradigms of comparative psychology and the threshold methods of psychophysics, has much in common with the study of shark (and other) repellents. In both fields, the effect of a noxious substance on behavior is scored.

Forty years ago (Tuve, 1947), shark repellents were tested with simple quantative methods. Although fairly sophisticated statistical methods were available at the time, the confluence of marine biology and comparative psychology was still in its infancy, and thus an available body of knowledge was overlooked. Today the situation for developing a bioassay for possible shark repellents is better. Because of advances in understanding shark behavior (Gruber and Myrberg, 1977) and maintenance (Gruber, 1980; Gruber and Keyes, 1981) and because of the development of comparative psychology (Stebbins, 1970), behavioral pharmacology, and behavioral toxicology (Weiss and Laties, 1969), a scientific framework is available to the marine biologist that simply was not formulated 40 years ago. We can now decide upon an experimental hypothesis relevant to shark repellents and design a behavioral bioassay that includes systematic selection of controls, treatments, experimental subjects, diagnostic behavioral responses, number of replications, and appropriate statistical methods (Mitchell, 1978).

Finally, the finding of a natural, apparently effective shark repellent by Clark (chapter 6 in this volume) has sparked new interest in shark repellency. Clark and Chao (1973) and Primor et al. (chapters 7 and 9 of this volume) have demonstrated an immediate and dramatic behavioral effect when certain sharks were exposed with relatively low concentrations of Pardachirus marmoratus secretion. Although their tests were relatively qualitative, they strongly suggested that this material rapidly and violently affects sharks in low concentrations. Thus, we have a chemical compound that can be tested in a scientific framework on a shark (the lemon shark) whose behavior can be well-controlled (Gruber and Schneiderman, 1975). In other words, we have all the necessary elements for an apparently valid behavioral bioassay.

Sharks as Experimental Subjects

Contrary to the general notion, certain sharks make exceptionally good laboratory subjects. One of the main stumbling blocks in past studies has been the difficulty of maintaining sharks in good physiological condition for long periods of time (Martini, 1978; Myrberg, 1976). More recently, however, a number of biologists have been able to keep certain species such as lemon (N. brevirostris), bull (C. leucas), nurse (Ginglymostoma cirratum), swell (Cephalosyllium retifer), horn (Heterodontus sp.), and several others (Gruber, 1980; Gruber and Keyes, 1981; Honn and Chavin, 1975; Nelson, personal communication). Closed aquarium systems (see Gruber, 1980) can provide controlled conditions in which these and other sharks feed well, grow rapidly, and appear to be healthy (see Figure 1).

Sharks have been used in behavioral studies, both ecological and physiological. Myrberg and Gruber (1974) have provided a limited ethogram and data on social behavior of the bonnethead shark (Sphyrna tiburo), and Nelson and his students (Finstad and Nelson, 1975; Nelson and Johnson, 1970) have completed long-term research on circadian rhythm in two shark species. Behavioral studies have involved conditioned responses of bull (C. leucas), lemon (N. brevirostris), and nurse (G. cirratum) sharks (e.g., Aronson et al., 1967; Clark, 1959; Graeber and Ebbesson, 1972; Gruber, 1967; Gruber and Myrberg, 1977). Thus, certain sharks are amenable to study by the usual methods of comparative psychology.

One of the best subjects to use in repellency studies is the lemon shark (N. brevirostris). This species is quite common in the tropics and easily captured. It survives the rigors of capture and can be kept in captivity for long periods throughout all stages of its life cycle. Thus, it is readily available for study. In addition, the lemon shark can be rapidly and reliably conditioned (Gruber and Myrberg, 1977; Gruber and Schneiderman, 1975) and subjected to experimental manipulations. Under test conditions, lemon sharks learn simple tasks at rates comparable to mammals (Gruber and Cohen, 1978). Finally, the lemon shark is a dangerous species (Garrick and Schultz, 1963; Randall, 1963, 1968), which has attacked humans (Baldridge, 1974). So this would be a valid species on which to base a behavioral bioassay for a repellent against shark attack.

Behavioral Studies

The typical bioassay uses physiological death as an objective measure of effect. The so-called LD_{50}--the dose which is fatal to 50% of the experimental population--is considered the lethal threshold in assay work. But LD_{50} is an inappropriate measure for repellency, which is, in fact, a type of behavior. What, then, is the appropriate and objective measure of repellency? Perhaps the best framework for finding an answer to this question is behavioral toxicology--the science that deals with the effects of nontherapeutic, nonphysiological chemicals on the behavior of intact subjects (Dews, 1978).

The field of behavioral toxicology arose in the late 1950s in response to Soviet research, which used Pavlovian conditioning techniques to establish clean air standards in the USSR (Ekel, 1974). Shortly thereafter Armstrong et al. (1963) published the first American behavioral bioassay of an industrial toxicant. Then the field expanded rapidly, drawing from the well-established methodologies of comparative psychology, psychopharmacology, and neuropsychology (Weiss and Laties, 1969). These methodologies are primarily protocols (i.e., experimental designs) for manipulating and measuring independent variables--in this case, repellent stimuli--so that their effects on behavior can be reliably determined (Lehner, 1979).

First, some general thoughts on the philosophy of experimental methods are appropriate here: Prior to embarking on an experimental study, the investigator should carefully consider and precisely formulate the experimental question(s) and then clearly state the objectives of the study and, more important, the research hypothesis. From these follow ex-

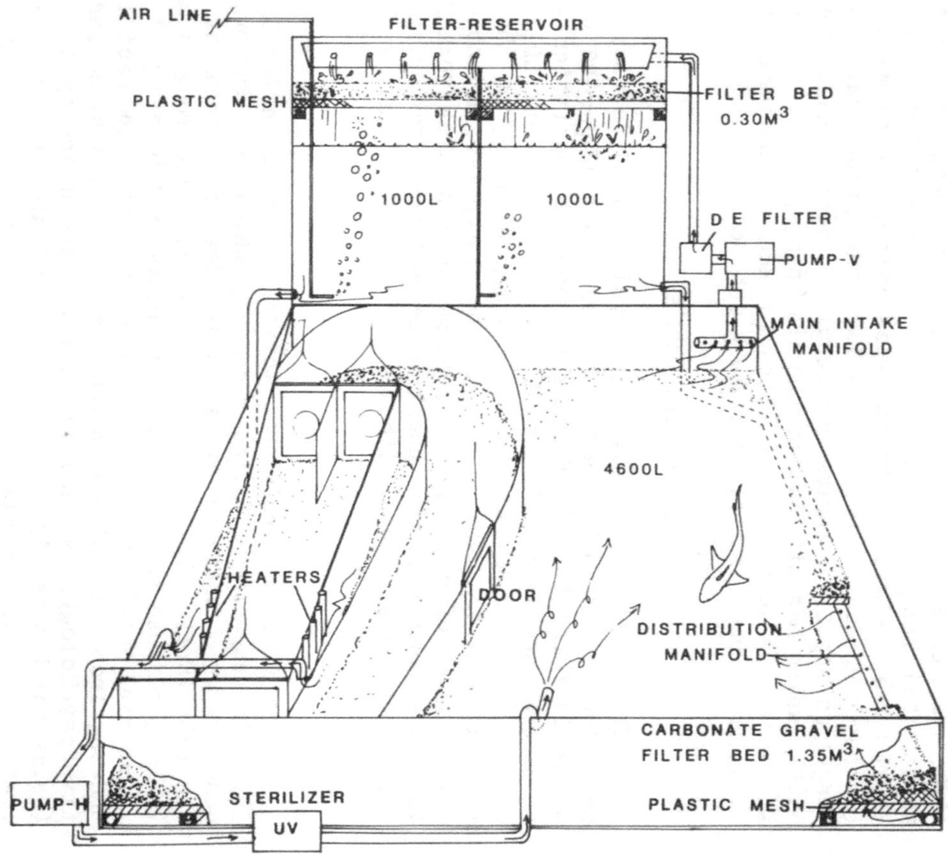
AIR LINE
FILTER-RESERVOIR
PLASTIC MESH
FILTER BED
0.30M3
1000L
1000L
D E FILTER
PUMP-V
MAIN INTAKE
MANIFOLD
4600L
HEATERS
DOOR
DISTRIBUTION
MANIFOLD
CARBONATE GRAVEL
FILTER BED 1.35M3
PLASTIC MESH
PUMP-H
STERILIZER
UV

Figure 1. (above) Perspective diagram of an aquarium and underwater "Y-maze" used in testing for color vision in the lemon shark. There are two circulation patterns in this 6600-liter system. For vertical circulation, water enters the intake manifold of pump "V" at mid-level, where it is pumped at 75 L/min through a diatomaceous (DE) filter and up to a filter reservoir. From there it flows by gravity through an 18 cm (in depth) carbonate filter bed into the reservoir, where it is aerated, and then flows into the main tank via distribution manifolds running the length of either side. These manifolds lie under a false bottom, so recycled water is channeled first under then up through the gravel on the tank bottom to return to the intake manifold, thus completing the vertical cycle. For the horizontal pattern, pump "H" takes water from areas of reduced circulation in the maze, pumps it through the sterilizer (UV) and finally out on the surface of the main tank. This sets up a slowly rotating current, which the sharks often swim against. (From Gruber, 1980, with permission) perimental design and statistical hypothesis. Even this beginning stage, if carefully considered, could take a great deal of time (Kerlinger, 1967). Lehner (1979) listed several criteria for choosing among the basic types of experimental design: (a) objective of the research; (b) hypothesis to be tested; (c) feasibility of gathering the raw data; (d) type of experimental design appropriate to the objectives; (e) level of precision; (f) level of statistical power; and (g) level of efficiency (i.e., cost-effectiveness of collecting the data). Once the experimental design, which is merely a tool for hypothesis testing, is chosen, another tool--namely statistical testing--must be employed because all biological data are inherently variable. Statistical tests enable us to tease out and identify the sources of variation in the data so that we may make decisions on whether to accept or reject statistical hypotheses. At some point, we must decide whether the difference between the control and treatment groups could have occurred by chance.

In sum, the research questions should be precisely formulated and transformed into a testable hypothesis. Then the experiment can be designed to include: selection of controls and treatments; selection of the diagnostic behavioral responses; number of replications; and statistical treatments. Permeating all of these considerations is the question of validity. How well does the chosen methodology and experimental subject actually answer the research question? How broadly applicable are the results? This is especially important in the laboratory evaluation of a shark repellent, which must eventually be tested on a number of shark species under widely varying field conditions.

We can divide behavioral methods into two groups: those that measure unlearned behavior and those that utilize learned (i.e., conditioned) responses. Natural behaviors run the spectrum from simple "reflexive" responses such as taxes and kineses (Vernberg et al., 1973) through higher order social behavior, including intraspecific communications (Todd et al., 1972). Although learning may have modified or given rise to these behavioral elements, such modifications were not induced by the experimenter (i.e., experimental subjects were not trained).

One advantage of using natural behavioral elements is that the assay can be easily and quickly performed. Responses of the unconditioned shark more closely reflect field behavior, and the significance of the unconditioned activity is usually easy to interpret. The major disadvantage is the high variability of an unlearned response. Such behaviors are usually dependent upon a variety of uncontrolled behavioral and environmental variables.

Another problem is learning per se. As an animal is exposed to an experimental situation, its experience will begin to affect its behavior through habituation, avoidance, or other learning phenomena. Thus, motivational levels will vary over the life of the experiment. This implies that the concentration of a repellent just adequate to repel a naive shark on the first trial may change on subsequent trials. One expensive way to avoid this problem is to use naive subjects for each test.

Olla (1974) divided behavioral bioassay of marine organisms into performance of individuals and interindividual behavior. He listed 24 categories of unlearned behavior in which effects of various noxious chemicals had been determined. Such a list could be expanded almost indefinitely, since both behavior and organisms are so variable. However, with regard to sharks--especially the lemon shark--three categories of natural behavior seem to lend themselves to bioassay. The first is that used in the World War II studies (Tuve, 1947) and in the highly suggestive behavioral studies of Eugenie Clark (chapter 6). At the onset, Clark's studies on the whitetip shark (Trianodon obesus) should be replicated on the lemon shark (N. brevirostris) and several other species to determine (a) if the violent repellent activity reported for the reef whitetip occurs in other species; and, more important, (b) if sharks never exposed to the Moses sole (P. marmoratus) in the field are repelled by its secretion. As pointed out by Baldridge (1976), we must first learn how universal Pardachirus repellency is and what role experience plays in shaping this behavior.

Lemon sharks feed very well in captivity, and their rates of food intake are relatively stable and predictable (Graeber, 1974; Longval et al., 1982). By observation under controlled conditions, one could replicate Clark's basic study to determine if Moses sole extract either inhibits feeding or actively repels lemon and other shark species. Once (and if) repellency is established, more quantitative studies should be done using feeding activity. For example, if the habituation problems are not significant, a dose-response curve could be established by interacting deprivation time (i.e., time during which food is withheld) with concentration of repellent material. For this test, food would be withheld for variable periods, then offered to naive lemon sharks. At a predetermined spatial location, a single bolus of repellent compound would be injected and the shark's response noted (see Figure 2), thus generating a dose-

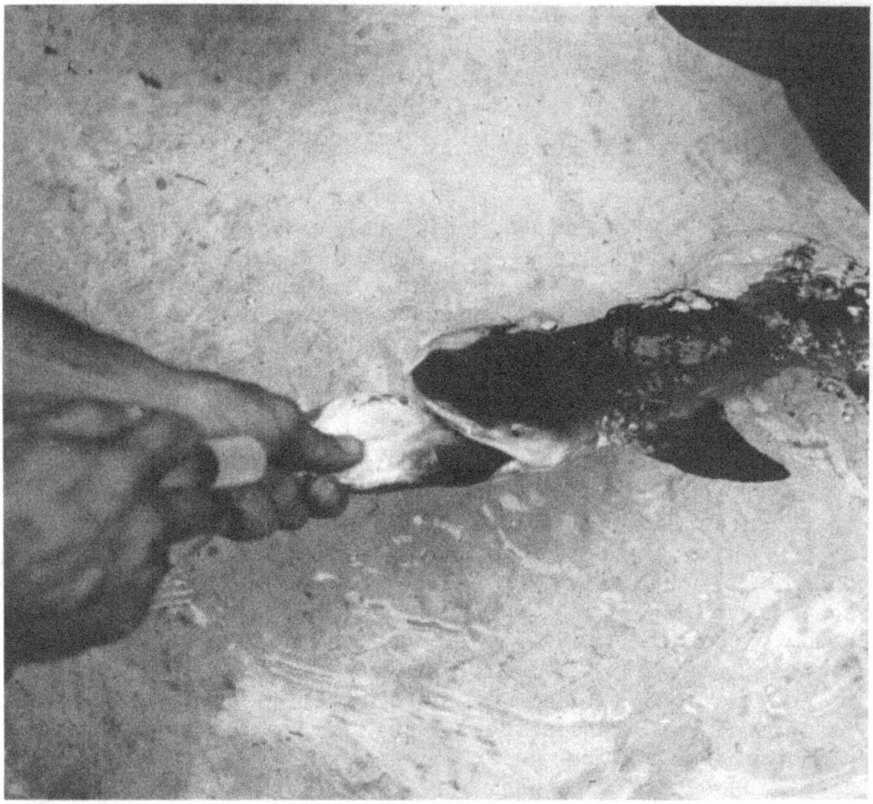

Figure 2. Preliminary feeding assay with the lemon shark (N. brevirostris) as a subject. Sharks were starved for 48 hours, then offered whole blue runner Caranx chrysos to which a syringe was attached. The syringe contained 4 mL of a specific concentration of a test substance. The photograph shows the shark at the moment the contents of the syringe are released into the mouth. Depending on the substance, the shark either continued to feed or was repelled.

response curve that would relate feeding rate to repellent concentration.

One of the most thoroughly investigated unconditioned behaviors in toxicity research involves the analysis of spontaneous motor activity. Rieter and MacPhail (1979) listed more than 130 such studies that primarily used the laboratory rat. The standard test compares movements of a control group in an activity chamber to that of a treated group. Such studies could easily be done with sharks. Kleerekoper et al. (1975) developed a sophisticated activity tank for the study of olfactory orientation in sharks. The system employed an array of photosensors embedded in the bottom of the tank and near-infrared light illuminating from above. The moving shadow of the shark signaled a computer, which printed a map of the animal's movements in the tank. Although sensory motor analysis is not particularly adapted to repellency testing per se, it could be used in combination with feeding studies to quantify changes in shark activity upon exposure to the repellent compound. Instead of noting that a particular shark was repelled, the computer would produce a hard copy of the shark's movement upon contact with the repellent substance. This would be a more objective method of describing behavior.

A third possibility of studying the activity of repellents on sharks is that of tonic immobility--a quiescent state of inactivity induced by restraining an animal in an inverted position (Carli, 1977). Tonic immobility can be induced in many kinds of animals including sharks (see Figure 3). As a quantifiable behavioral system, it is stereotyped, resistant to habituation (at least in sharks), and has a definite endpoint: spontaneous recovery. A possible test method would involve inducing tonic immobility and applying different repellent substances in varying concentration to different parts of the shark's body, especially the bucal cavity and sensory end organs. The minimum concentration of test material that terminated the immobile state would represent the threshold value. Thus, possible sites of activity and dose-response curves could be evaluated. This test might represent a rapid bioassay for comparing the relative strength of a repellent such as pardaxin (see Primor et al., chapter 9) to other repellent-activating compounds. Since tonic immobility can be induced in large sharks, one could make a quick evaluation of a repellent substance in the field.

The development of assay methods in behavioral toxicology has focused on established bahavioral procedures (Mello, 1975). Bioassay based on learned behaviors has assumed a greater importance, probably because of the vast

Figure 3. Photograph of a lemon shark (N. brevirostris) under tonic immobility. Inverting the shark and restraining it in the position shown for 30 sec induces a transient state of immobility, which can last for more than 30 min. The tonic immobility response was developed as a bioassay by Gruber and Zlotkin (see Gruber, 1982).

theoretical framework provided by both experimental psychology and behavioral pharmacology. Moreover, conditioned responses are more stable, more predictable, and less variable than unlearned responses. The sequential nature of conditioning paradigms allows for the quantitative analysis of behavior into separate measureable dimensions. Thus, a small effect might go entirely unnoticed in an evaluation of gross spontaneous behavior such as sensory motor analysis. The implication is that conditioned response methods may provide tests of greater sensitivity. Finally, the production of a highly stable baseline of behavior characteristic of conditioning is a prerequisite for precise measurement of behavioral disruption, particularly since we must account for the sources of variation. Statistical tests developed over the past 50 years of experimental psychology represent a rich resource for such hypothesis testing in behavioral toxicology and repellency studies.

There are two basic learning paradigms in comparative psychology: Pavlovian (classical) conditioning and operant (instrumental) conditioning (Kling and Riggs, 1971). In classical conditioning, an unconditioned stimulus such as mild electric shock reliably elicits an unconditioned response. For example, an animal restrained in a cage might flex or withdraw its leg when a small electric shock is applied to the foot. A neutral or conditioned stimulus such as an audible tone can be paired in a specific temporal relation with the unconditional stimulus. In the case of standard classical conditioning, the tone would precede the shock; after a variable period of such pairings, the tone alone will elicit flexation of the leg (leg withdrawal in this case is known as a classically conditioned reflex). In this paradigm, the experimenter, not the animal, controls the conditioning through the establishment of conditioned-stimulus/unconditioned-stimulus temporal relation. No matter what the animal does, it cannot escape the shock.

In the case of instrumental conditioning, a switch might be connected to the circuit so that when the animal withdraws its leg, it avoids the shock. Thus the sounding of the tone might signal the onset of shock, but--unlike classical conditioning--the behavior of the animal determines the outcome. Actually, instrumental behavior can be modified in endless ways by appropriate manipulation of its consequences.

Lemon sharks have been excellent experimental subjects for both classical and instrumental conditioning (Gruber and Myrberg, 1977), as shown in Figure 4. Training consisted of trials in which a neutral flash of light (conditioned stimulus) was paired with a low-voltage electrical shock (uncondi-

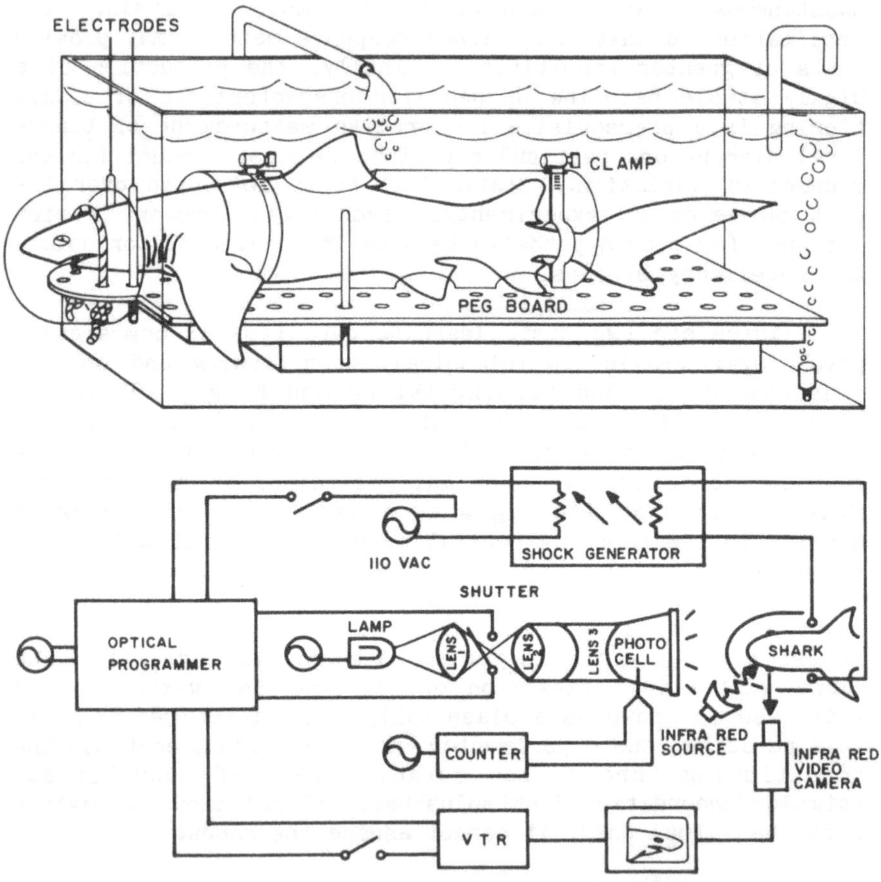

Figure 4. Apparatus used to train lemon sharks by aversive classical conditioning. Upper sketch shows how the shark was restrained and fixed in position, head in a clear acrylic dome. Lower diagram shows the sequencing devices that permitted automatic presentation of conditioning trials. (From Gruber and Schneiderman, 1975, with permission)

tioned stimulus). At the onset, the flash of light did not elicit an observable response, whereas the 100 ms shock always evoked a reflexive movement (unconditioned response) of the third eyelid or nictitating membrane. During training, 100 trials were given each day; the interval between trials was variable, but averaged 30 seconds. With a 10-minute adaptation period prior to the trials, a single session lasted about one hour. After about 30 pairings in the first session, the subject began to blink before the onset of the shock; that is, the shark anticipated the electric shock and responded with a classically conditioned response by moving its nictitating membrane. By the third session, the shark was responding with nearly 100% accuracy (see Figure 5). When cats and rabbits are conditioned in a similar way (i.e., light as the conditioned stimulus and movement of the nictitating membrane as the conditioned response), both species require several hundred more pairings than the lemon shark to form the conditioned response. Thus, in this paradigm the shark learns much more rapidly and predictably than cats or rabbits (Gruber and Myrberg, 1977; Gruber and Schneiderman, 1975).

There are at least three ways classical conditioning could be related to toxicology studies: Learning curves, extinction curves, and sensory deficits could be compared in treated and control groups. However, for repellency studies, this method is perhaps not as instructive as operant conditioning, because the shark is restrained and spatially fixed in classical conditioning trials. Yet classical conditioning is advantageous for measuring thresholds, because it facilitates careful evaluation of the stimulus field about an immobile animal. Sharks could be conditioned by using the aversive properties of a brief exposure to a repellent compound as the unconditioned stimulus. Once a conditioned response were established, the experimenter could reduce the concentration of the compound in small steps over a number of trials until a further reduction failed to elicit a conditioned response. This psychophysical technique, known as the method of limits (Stebbins, 1970), is often used to measure sensory thresholds. It thus evaluates the minimum concentration of a repellent necessary to elicit a conditioned response (i.e., threshold concentration).

The other general paradigm of associative learning, operant or instrumental conditioning, is probably more adaptable to repellency studies because of the ease in training whole animal movements. Operant conditioning of sharks has involved techniques ranging from bar pressing (Clark, 1959) to entering a goal box (Aronson et al., 1967) to using a single-choice "Y-maze" (Graeber, 1978; Gruber and Myrberg,

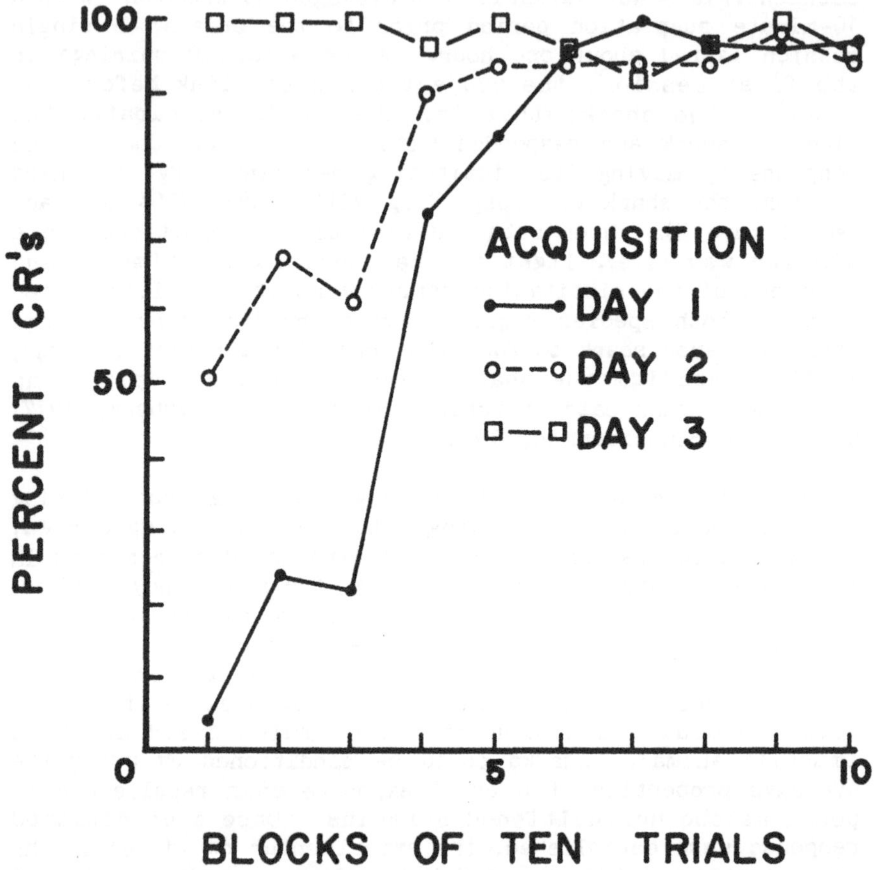

Figure 5. Course of acquisition of a classically conditioned movement of the nictitating membrane of the lemon shark (N. brevirostris). Training consisted of pairing a flash of light with a low-voltage electric shock 100 times per day (i.e., 10 blocks of 10 trials). Three days of training are shown. Note that the sharks reached nearly 100% conditioned responses by the 60th trial of the first day. (From Gruber and Schneiderman, 1975, with permission)

1977). In at least one study (see Figure 6), nurse sharks (G. cirratum) learned at essentially the same rate as teleosts and rats trained under nearly identical stimulus conditions.

I have trained lemon sharks through appetitive discrimination conditioning to choose one of a pair of doors in an underwater "Y-maze." The sharks swim down an alley to a choice point; by choosing the most intensely illuminated door (see Figure 7), they may swim through and obtain a food reward just behind the correct door. If they choose incorrectly, the door remains shut and the sharks must laboriously back out and swim through the correct door. Such a trial goes unreinforced. Thus, lemon sharks can be conditioned in about 15 sessions or 300 trials (see Figure 8), a rate which, in a simple brightness discrimination, compares favorably with that of some mammals.

Clearly, then, lemon and other shark species make relatively good subjects for controlled experiments using associative learning as a behavioral criterion. Thus, one major confounding variable (i.e., motivational state of subject) can be reasonably well controlled. Since shark attacks may be motivated by factors other than feeding, food deprivation is one way to control motivation. According to Bartoshuk (1971), the use of food deprivation in motivation control does not limit the behavior to a get-food mechanism. For example, many studies have shown that food deprivation can facilitate a wide variety of response tendencies besides those regulating eating (Miller, 1957). Thus, the use of food as a reinforcer in instrumental learning is quite different and perhaps more valid than the use of food with a hungry unconditioned shark.

A number of different behavioral tests might use an aversive chemical stimulus either as a reinforcer or as a test substance. For example, sharks could be trained to enter a chamber and press a lever to obtain food reward (Aronson et al., 1967; Clark, 1959). After a fixed number of hours of food deprivation, the conditioned shark should enter a "goal box" and press the lever at a known rate (bahavioral baseline). At this point, minute quantities of the repellent substance could be introduced at the response lever, and effects on the shark's behavior would be recorded. This method permits control over both the animal's position in space and its behavior at the time the test substance is introduced. It is a vast improvement over simple unconditioned feeding experiments in which hungry sharks are fed in the presence of repellent substances (Tuve, 1947).

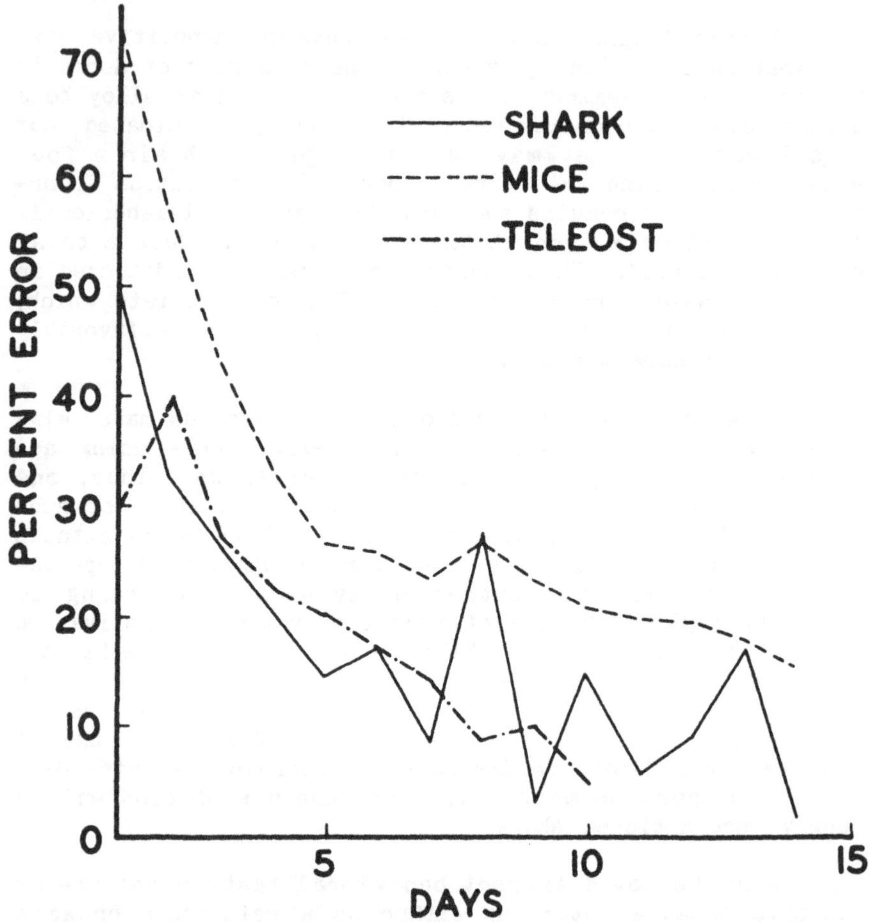

Figure 6. Learning curves for three vertebrates on a brightness discrimination problem. The stimulus conditions and responses were very similar for each subject. Although the nurse shark (G. cirratum) was somewhat more erratic in learning than the other subjects, its rate of learning was about the same. (After Aronson, from Gruber and Myrberg, 1977, with permission)

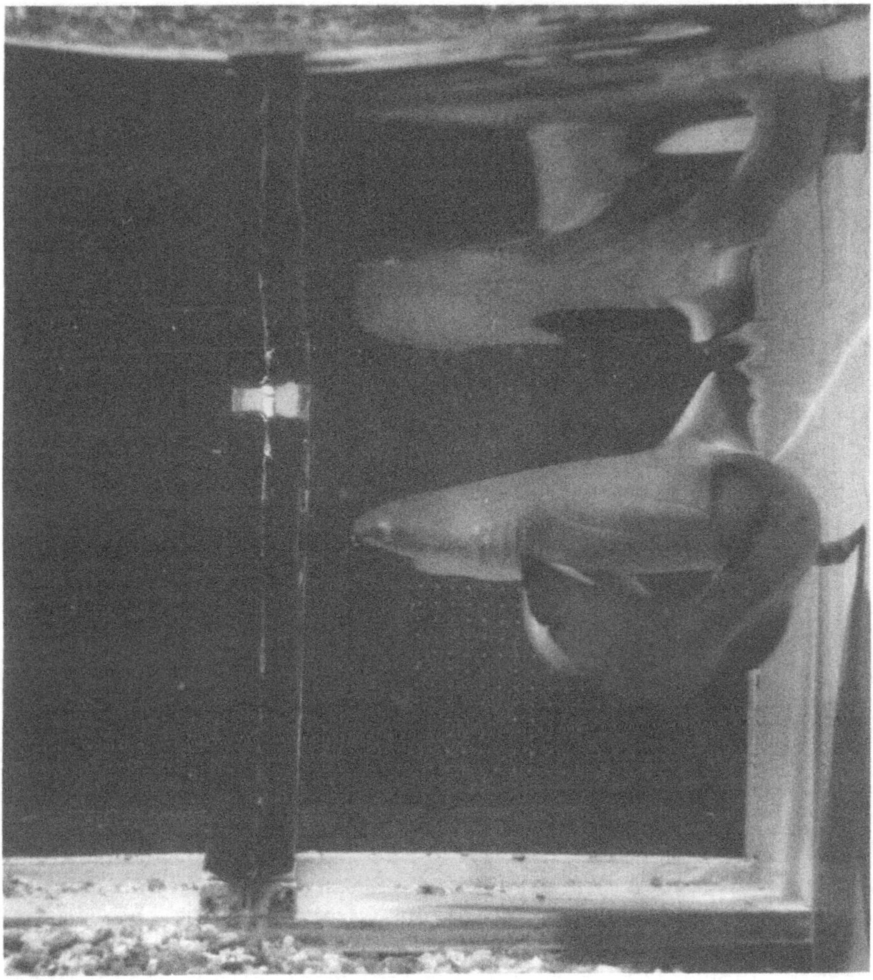

Figure 7. Underwater photograph of a lemon shark at the choice point of a "T-maze." The shark is choosing the right-hand door of a pair of doors, each illuminated by an array of optical fibers connected to a light source. For choosing the brighter of the pair, the shark earns a food reward in the reinforcement area behind the doors. (Photo by C. Bry; from Gruber and Myrberg, 1977 with permission)

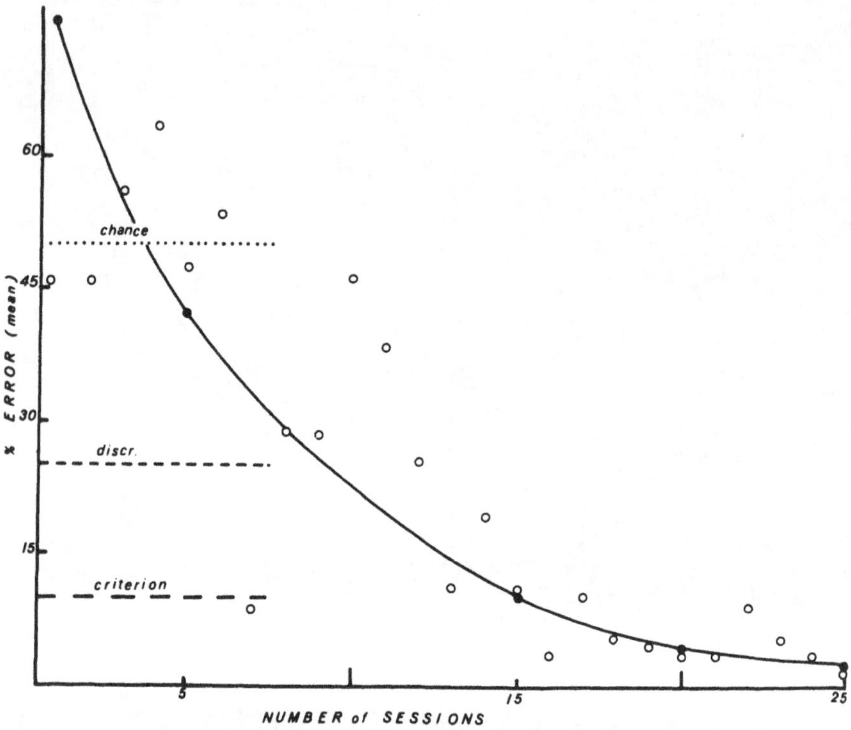

Figure 8. Learning curve of lemon shark on the brightness discrimination task shown in Figure 7. Open circles represent mean % errors for six animals; closed circles were calculated from a standard curve-fitting procedure. Learning is signaled by reduction in errors, i.e., choosing the dimmer door. Chance refers to random choice, i.e., the 50% correct level. Discr. refers to the 75% correct limit of discrimination usually allowable in psychophysical testing, whereas criterion refers to the arbitrary 90% correct level chosen in this study. (From Gruber and Myrberg, 1977, with permission)

Concluding Remarks

Recent developments in behavioral toxicology, experimental psychology, and maintenance of sharks should facilitate development of a bioassay for chemical repellents with sharks as subjects. Many of the speculative concepts outlined in this paper were recently confirmed in my laboratory in collaboration with E. Zlotkin (Gruber and Zlotkin, 1982). Zlotkin provided substances that were tested on captive lemon sharks. First, we ran a simple feeding trial: Food-deprived sharks were allowed to grasp the head of a whole fish in their mouth; then a bolus of test material was injected into their bucal cavity via a syringe sutured in the bait fish. We noted the outcome of the exposure on the continuation of the shark's attack. With the most effective repellents, the actively feeding shark immediately ceased the attack and rapidly left the feeding station.

A second, and apparently more reliable, bioassay involved the use of tonic immobility: We injected a known concentration of test substance into the bucal cavity of a shark placed under tonic immobility. The concentration that terminated the immobile state (i.e., causing the shark to right itself) was considered the threshold. There was agreement between tonic immobility and feeding trials, and we were able to identify three substances that, in moderately low concentration ($\mu g{\cdot}mL^{-1}$), clearly repelled actively feeding sharks. Thus, in less than three weeks, we developed two rapid assay methods and surveyed eight different substances, including the crude secretion of P. marmoratus. With further refinement of these methods and development of an instrumentally conditioned bioassay, we should have the raw material to screen large numbers of substances effectively in a relatively short time.

Acknowledgments

I wish to express my gratitude to William Lee Hearn for his help in preparing the original lecture for the 1981 AAAS Meeting; Hearn helped develop many of the arguments presented herein. I also wish to acknowledge Eliahu Zlotkin for his belief in my behavioral schemes and ultimately for collaborating with me. The preliminary repellent study of Zlotkin and Gruber was supported by a contract between the University of Miami and the Office of Naval Research.

References Cited

Armstrong, R.D., J.E. Leach, P.R. Belluscio, E.A. Maynard, H.C. Hodge, and J.K. Scott. 1963. Behavioral changes in the pigeon following inhalation of mercury vapor. Am. Ind. Hyg. Assn. J. 24:366-375.

Aronson, L.R., F.R. Aronson, and E. Clark. 1967. Instrumental conditioning and light-dark discrimination in young nurse sharks. Bull. Mar. Sci. 17:249-256.

Baldridge, D. 1974. Shark attack: a program of data reduction and analysis. Contrib. Mote Mar. Lab. 1:1-98.

_____. 1976. A reminder of the impracticability of chemical shark repellents. Page 18 in W. Seaman, Jr., ed. Sharks and Man: A Perspective. Florida Sea Grant Report No. 10, Gainesville, Fla.

Bartoshuk, A.K. 1971. Motivation. Pages 793-845 in J.W. Kling and L.A. Riggs, eds. Woodworth and Schlosberg's Experimental Psychology. Holt, Rinehart, and Winston, New York.

Carli, G. 1977. Animal hypnosis in the rabbit. Psychol. Rec. 27:123-143.

Clark, E. 1959. Instrumental conditioning in lemon sharks. Science 136:217-218.

Clark, E. and S. Chao. 1973. A toxic secretion from the Red Sea flatfish Pardachirus marmoratus (Lacepede). Bull. Sea Fish. Res. Stn. (Haifa) 60:53-56.

Dews, P.B. 1978. Epistemology of screening for behavioral toxicity. Environ. Health. Perspect. 26:37-42.

Ekel, G.J. 1974. Soviet behavioral toxicology research. Pages 432-440 in C. Xintaras, ed. Behavioral Toxicology. U.S. Government Printing Office, NIOSH, Washington, D.C.

Finstad, W.D., and D.R. Nelson. 1975. Circadian activity rhythm in the horn shark, Heterodontus francisci: effect of light intensity. Bull. Southern Calif. Acad. Sci. 74:20-26.

Garrick, J.A., and L.P. Schultz. 1963. A guide to the kinds of potentially dangerous sharks. Pages 3-60 in P.W. Gilbert, ed. Sharks and Survival. D.C. Heath and Co., Boston.

Graeber, R.C. 1974. Food intake patterns in captive juvenile lemon sharks, Negaprion brevirostris. Copeia

1974:554-556.

_____. 1978. Behavioral studies correlated with central nervous system integration of vision in sharks. Pages 195-225 in E.S. Hodgson and R.F. Mathewson, eds. Sensory Biology of Sharks, Skates and Rays. Office of Naval Research, Arlington, Va.

Graeber, R.C., and S.O. Ebbesson. 1972. Visual discrimination learning in normal and tecta-ablated nurse sharks (Ginglymostoma cirratum). J. Comp. Biochem. Physiol. 42A:131-139.

Gruber, S.H. 1967. A behavioral measurement of dark adaptation in the lemon shark, Negaprion brevirostris. Pages 479-490 in P.W. Gilbert, R.F. Mathewson, and D.P. Rall, eds. Sharks, Skates and Rays. Johns Hopkins Press, Baltimore.

_____. 1980. Keeping sharks in captivity. J. Aquaricult. 1:6-14.

_____. 1982. Shark repellents: perspectives for the future. Oceanus 24(4):72-76.

Gruber, S.H., and J.L. Cohen. 1978. Visual system of the elasmobranchs. State of the art 1960-1975. Pages 11-106 in E.S. Hodgson and R.F. Mathewson, eds. Sensory Biology of Sharks, Skates and Rays. Office of Naval Research, Arlington, Va.

Gruber, S.H., and R.M. Keyes. 1981. Keeping sharks for research. Pages 373-402 in A. Hawkins, ed. Aquarium Systems. Academic Press, London.

Gruber, S.H., and A.A. Myrberg, Jr. 1977. Approaches to the study of the behavior of sharks. Am. Zool. 17:471-486.

Gruber, S.H., and N. Schneiderman. 1975. Classical conditioning of the nictitating membrane response of the lemon shark (Negaprion brevirostris). Behav. Res. Meth. Inst. 7:430-434.

Gruber, S.H., and E. Zlotkin. 1982. Bioassay of surfactants as shark repellents. Nav. Res. Rev. 32:1-10.

Honn, K.V., and W. Chavin. 1975. Prototype design for a closed marine system employing quaternary water processing. Mar. Biol. 31:293-298.

Kerlinger, F.N. 1967. Foundations of Behavioral Research. Holt, Rinehart and Winston, New York.

Kleerekoper, H., D. Gruber, and J. Matis. 1975. Accuracy of localization of a chemical stimulus in flowing and stagnant water by the nurse shark Ginglymostoma cirratum. J. Comp. Physiol. 86:27-31.

Kling, J.W., and L.A. Riggs. 1971. Woodworth and Schlosberg's Experimental Psychology. Holt, Rinehart and Winston, New York.

Lehner, P.N. 1979. Handbook of Ethological Methods. Garland STPM Press, New York.

Longval, M.J., R. Warner, and S.H. Gruber. 1982. Food intake of the lemon shark, Negaprion brevirostris under contolled conditions. Fla. Sci. 45:25-33.

Martini, F.H. 1978. The effects of fasting confinement on Squalus acanthias. Pages 609-646 in E.S. Hodgson and R.F. Mathewson, eds. Sensory Biology of Sharks, Skates and Rays. Office of Naval Research, Arlington, Va.

Mello, N.K. 1975. Behavioral toxicology: a developing discipline. Fed. Proc. 34:1832-1834.

Miller, N.E. 1957. Experiments on motivation. Science 65: 318-333.

Mitchell, C.L. 1978. Target organ toxicity: nervous system. Environ. Health Perspec. 26:3-4.

Myrberg, A.A., Jr. 1976. Behavior of sharks--a continuing enigma. Nav. Res. Rev. 29:1-11.

Myrberg, A.A., Jr., and S.H. Gruber. 1974. The behavior of the bonnethead shark, Sphyrna tiburo. Copeia 1974: 358-374.

Nelson, D.R., and R.H. Johnson. 1970. Diel activity rhythms in the nocturnal bottom dwelling sharks Heterodontus francisci and Cephaloscyllium ventriosum. Copeia 1970:732-739.

Olla, B. 1974. Behavioral measures of environmental stress. Pages 1-31 in G.V. Cox, chm. Proceedings of a Workshop on Marine Bioassays. Marine Technology Society, Washington, D.C.

Randall, J.E. 1963. Dangerous sharks of the western Atlantic. Pages 339-361 in P.W. Gilbert, ed. Sharks and Survival. D.C. Heath and Co., Boston.

_____. 1968. Caribbean Reef Fishes. TFH Publishing Inc., Jersey City.

Reiter, L.W., and R.C. MacPhail. 1979. Motoractivity: a survey of methods with potential use in toxicity testing. Neurobehav. Toxicol. 1(suppl. 1):53-66.

Stebbins, W.C. 1970. Animal Psychophysics: The Design and Conduct of Sensory Experiments. Plenum Press, New York.

Todd, J.H., S. Engstrom, S. Jacobson, and W.O. McLarney. 1972. An introduction to environmental ethology: preliminary comparison of sublethal thermal and oil stresses on the social behavior of lobsters and fishes from a freshwater and marine ecosystem. Unpublished. U.S. Atomic Energy Comm. Prog. Rept. AT(11-1) 3567.

Tuve, R.L. 1947. Technology of the U.S. Navy "Shark Chaser." U.S. Navy Inst. Proc. 73:522-527.

Vernberg, W.B., P. DeCoursey, and W.J. Padgett. 1973. Synergistic effect of environmental variables on larvae of Uca pugilator (Bosc.). Mar. Biol. 22:307-312.

Weiss, B., and V.G. Laties. 1969. Behavioral pharmacology and toxicology. Annu. Rev. Pharmacol. 9:297-342.

Joseph Bonaventura, Celia Bonaventura

5. Flatfish, Fireflies, Sharks: Behavior Modification Induced by Natural Repellents

Like buried treasures, the resources of the oceans have beckoned to the adventurous from immemorial times. Finding and exploiting these treasures, though, have always been linked to problems associated with the unpleasant possibility of shark attack. One purpose of this symposium is to see if the problem of finding a repellent for sharks is solvable in some practical and attainable way.

In the solution of matters that involve science, there are three classes of problems: those whose solution depends on only small numbers of accessible bits of knowledge (like optimizing a mechanical system); those whose solution depends on necessary bits of knowledge that are as yet unavailable (like curing the common cold); and those whose solution involves great numbers of parameters (like improving the world's economy). The shark repellent problem falls in the second category; it is solvable, but not all of the relevant information is at hand.

Sharks are perhaps the most well known of the ocean's predatory species. To pose the question of whether an effective shark repellent can be developed is to ask if man can fit himself to life in the underwater environment, as has been done by many natural inhabitants of the sea. If we ask how sea creatures have dealt with the problems associated with the predatory habits of sharks, a survey would show that the evolutionary process has led to species that survive by being faster, stronger, smarter, camouflaged, or mechanically or chemically distasteful.

The species Homo sapiens, terrestrial by nature, is not faster, stronger, or easily camouflaged when in the sea. Although we can perhaps claim a higher level of intelligence, this feature has not been reassuring to lone swimmers who have seen evidence of nearby sharks. The mental superiority

Table 1. Typical sensory systems of the environment.

Sense	Medium	Stimulus
The five classical senses		
Sight	Photons	
Sound	Phonons	
Touch	Mechanical Displacement, Thermal Motion	
Taste	Glucose	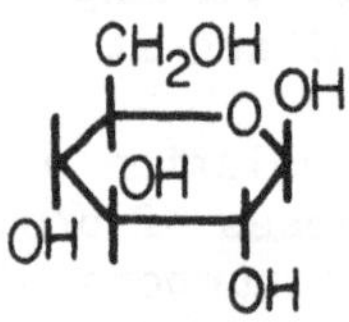
Smell	Geraniol 	Menthol 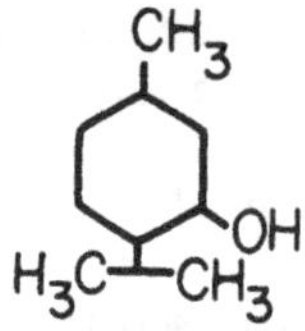
Additional nonclassical senses		
Electric	Electrons e^-	
Magnetic		
Light Polarization	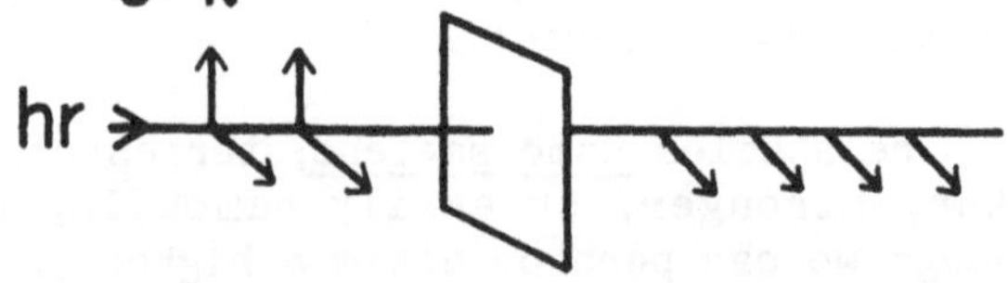	

we claim has not been directed by evolutionary pressures toward learning how to deal with these powerful marine predators.

With these considerations, then, this symposium on shark repellents was scheduled, to pool the thoughts and experiments of many interested and experienced individuals on the problem. To this end, this paper presents our bias toward basing a repellent on the shark's own chemical recognition system. First, we give our definition of chemical recognition, and then we describe how this conceptual framework can be used in a biochemical approach toward developing an effective shark repellent. Finally, we discuss how a biochemical approach can lead to new concepts for delivery systems for shark repellents.

Chemical Recognition

Organisms perceive the quality of their environment by sensory systems whose output is amplified and used to direct behavior. Repellents, as well as attractants, must act by stimulating a sensory system. Of the five senses commonly used by humans, many terrestrial organisms possess similar sensory systems based on visual, auditory, tactile, taste, or olfactory input. Table 1 indicates the most common sensory systems of higher organisms, as well as less common systems, such as those responsive to electrical or magnetic stimulation. Stimuli such as light polarization or the direction of fluid flow may also provide for communication by stimulation of a sensory system.

The sensory organs of sharks are shown in Figure 1. Figure 2 shows approximate distances at which sharks perceive stimuli of various types (see Hodgson and Mathewson, 1978). Communication may be achieved by directing single or complex stimuli to one or more of the animal's sensory receptors. The content of the "message" would, of course, depend on the nature of the stimulant (see Hodgson and Mathewson, 1978, for a review of sensory biology of elasmobranchs).

Of these sensory systems, none is more readily demonstrated in sharks than that related to detection of food. To the consternation of those who would like to camouflage man in the water, sharks seem to have a well-developed chemical recognition system. Cloudy or opaque water does not prevent them from detecting (and ingesting) a tasty morsel. For those who have observed the behavior of predatory sharks, the pattern of food detection, circling, and attack is memorable and distinctive. Thus, a repellent stimulus could be focused on the food recognition system.

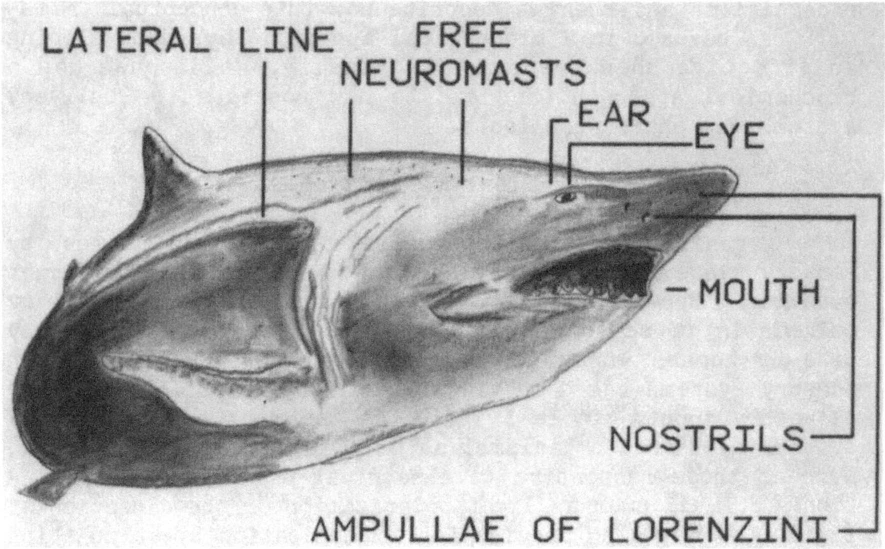

Figure 1. Some types of sensory perception systems found in sharks.

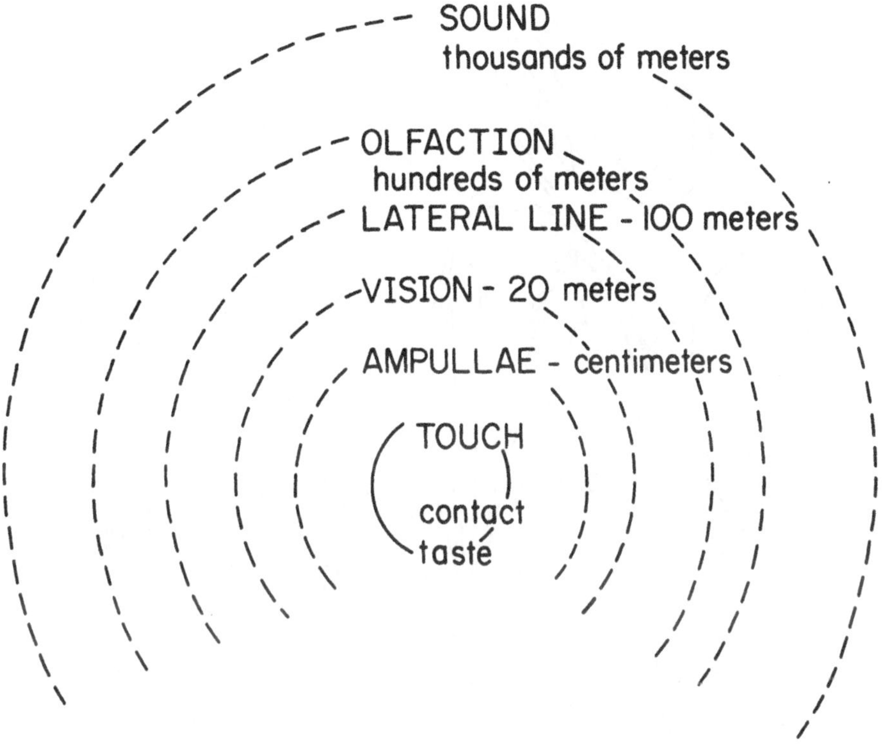

Figure 2. Generalized scheme of the reception of stimuli in sharks.

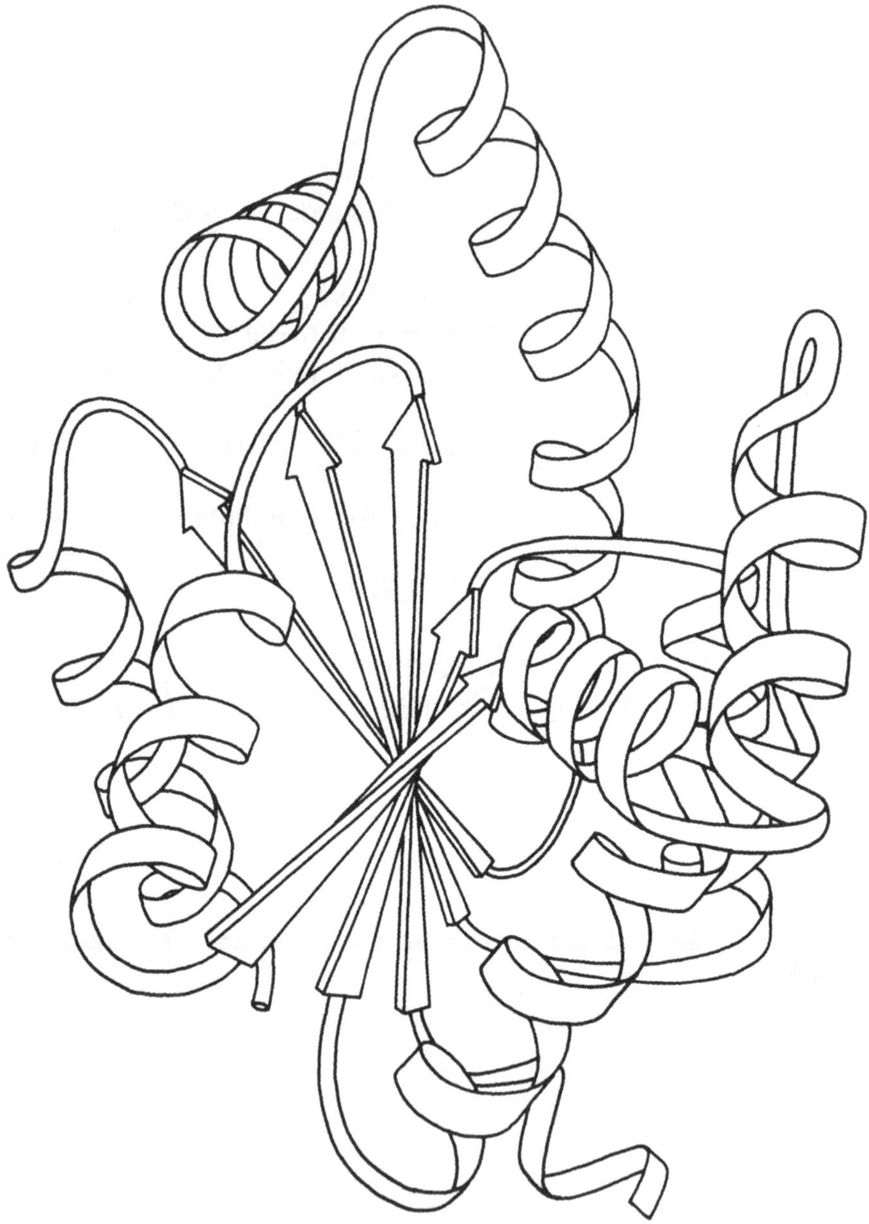

Figure 3. Schematic representation of the enzyme adenylate kinase, which uses ATP to phosphorylate AMP to ADP. It, therefore, has chemical recognition sites for AMP (the substrate) and ATP (the source of the "high energy" phosphate that phosphorylates the substrate). The drawing is used by permission of Jane Richardson, Little River Institute, Bahama, North Carolina.

Food detection is an example of a more general phenomenon, which we call the chemical recognition process. Naturally, other senses are involved as well in food detection. Although it is often overlooked, chemical recognition underlies many metabolically important events.

Inorganic materials can form complexes whose stability is variable and highly dependent upon the mutual affinity of the interactive materials. The same is true of biochemical complexes, but the degree of specificity of complex formation that can be achieved is very high in comparison. Figure 3 depicts the structure of a protein involved in the catalysis of a chemical reaction. The standard way to describe the specificity of the catalysis has been to invoke the image of a "lock and key" arrangement between the enzyme and its substrate. A further suggestion states that conformational changes in both the enzyme and substrate are responsible for the specificity of the fit which triggers the catalysis. Thus, the constituents in a biological interaction may play dynamic roles in the catalytic reaction, and need not be considered as rigid or fixed elements.

This view of the chemical recognition process emphasizes the importance of the time course of complex formation and dissolution. One example is the enhancement or inhibition of a reaction by the formation of an enzyme-cofactor or inhibitor complex. Figure 4 depicts the fit between a protein and a cofactor by showing a positive charge cluster of the hemoglobin molecule, where organic anions like 2,3-diphosphoglycerate (shown in the figure), ATP, and inositol hexaphosphate are bound (Arnone, 1972).

Enzymatic reactions can often be inhibited by substrate analogues, that is, chemicals that are recognized by the enzyme because they possess some of the groups or "recognition elements" that are used in substrate recognition. Thus, there may be an analogue to a substance that is favorably recognized by a shark but which carries a repellent message. Recognition elements are well-known for their specificity in living systems. The antigen-antibody response is one of the more well-developed examples. This system is so specific that antibodies can discriminate two hemoglobin molecules that differ from each other by a single amino acid interchange in the protein structure (this change leads to alteration of a single amino acid in the 287 amino acids of an $\alpha\beta$ hemoglobin dimer).

The high selectivity of the chemical recognition process enables a sensory system to detect amazingly low concentrations of a stimulus, such as the attraction of insects by

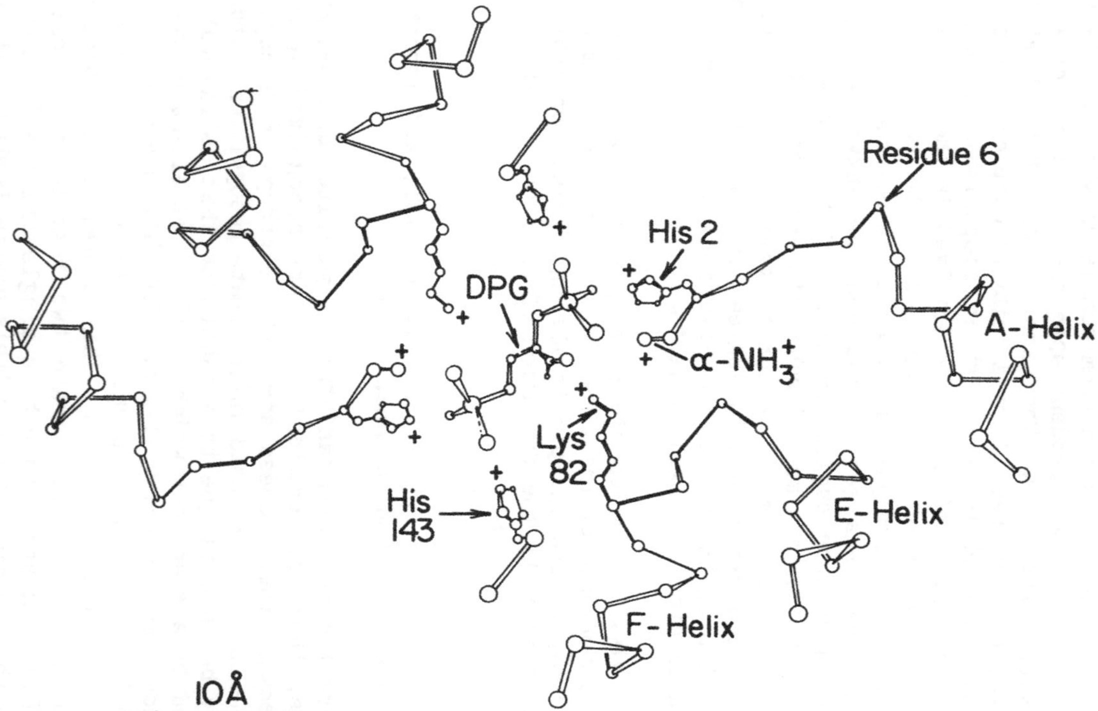

Figure 4. Schematic representation of the recognition, or binding, site for hemoglobin's allosteric modifier 2,3-diphosphoglycerate. A cluster of positively charged amino acids from two beta chains of the hemoglobin molecule provide a sterically complementary binding site for the diphosphoglycerate molecules.

pheromones. Attraction of sharks by potential food sources are also examples of effective stimulation at low concentration. Directional action is possible in these cases by way of the ability of the sensing system to monitor the chemical gradient of the stimulus.

Chemical recognition and detection of chemical gradients are involved in many behavioral responses of organisms. As we appreciate the great variety of these phenomena, we will improve our techniques for their analysis and detection, as entomologists have done with their studies on sex pheromones, alarm pheromones, and trail pheromones. The alarm pheromones, in particular, imply repellent characteristics: Many stinging insects discharge them along with the sting, thus "tagging" the opponent as an aggressor; the diffusion of the pheromonal substance then acts as a warning to other insects. For sharks, such a repellent will probably not be found in a predatory aggressor, since none is known. Nevertheless, there may be natural shark repellents. This seems to be the case in studies of a Red Sea flatfish, the Moses sole (*Pardachirus marmoratus*), and the sea hare (*Aplysia brasiliana*) (Hodgson and Mathewson, 1978).

Flatfish and Sharks: Protein-Mediated Chemical Communication

Have organisms developed natural repellents that utilize the shark's system for chemical recognition? Although further experimentation will be beneficial, a milky secretion from flatfish of the genus *Pardachirus* seems to be a natural repellent that protects the fish from predation by sharks (see Clark, 1974; Clark and Chao, 1973; Clark and George, 1979). Because of the considerable interest in this secretion, other reports in this symposium discuss its chemical nature and the field studies which have been done with it (see chapters 6,7,8, and 9).

Our involvement with the flatfish began when, in the course of creative research management, an Office of Naval Research program manager arranged for one of his contractors to tour several marine research laboratories. At our laboratory, we listened with enthusiasm to Naftali Primor's seminar on the biochemical aspects of the milky secretion of the Red Sea flatfish, *Pardachirus marmoratus* (see Parness and Zlotkin, 1976; Primor and Zlotkin, 1975; Primor et al., 1978, 1980). We learned that the prime component of the secretion, pardaxin, has not only an apparent repellent effect on sharks, but also a relatively rapid and lethal effect on small teleost fish (killifish, *Fundulus heteroclitus*). Teleost gill preparations show an inhibition of Na^+-K^+-ATPase

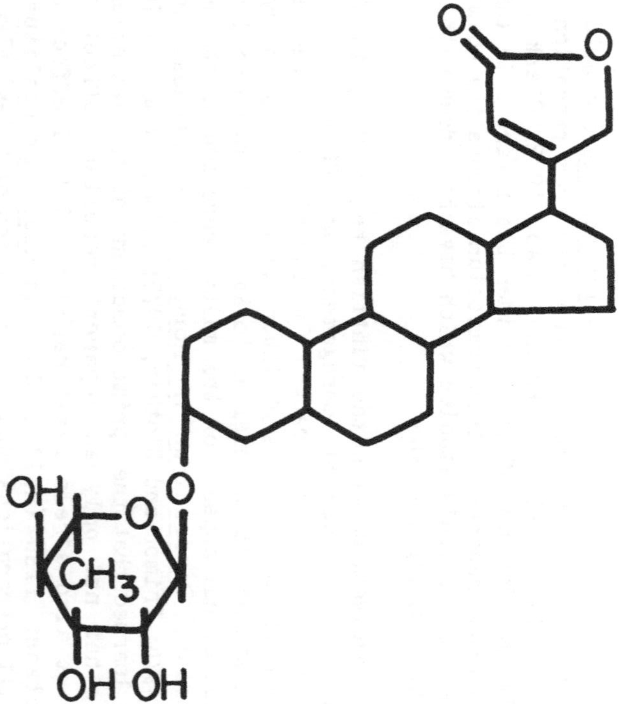

Figure 5. Diagram of the chemical structure of ouabain, a potent inhibitor of Na^{+}-K^{+}-ATPases.

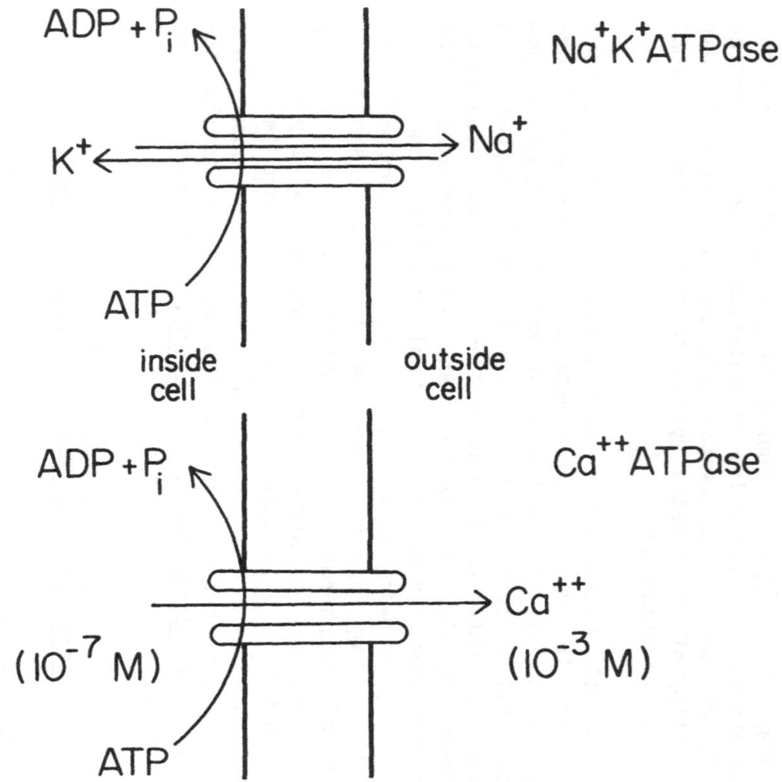

Figure 6. Schematic representations of Na^{+}-K^{+}-ATPase (top) and Ca^{++}-ATPase (bottom).

activity at very low toxin levels (see Primor et al., 1980).

Clearly, further insight into the mode of pardaxin action could be gained by controlled laboratory tests with sharks. At that time we had a colony of small sharks whose characteristics made them suitable for laboratory tests. These Atlantic sharpnose sharks (*Rhizoprinodon terraenovae*) are small in size but similar in habit to large predatory species. (We were maintaining the colony to study heavy metal uptake in sharks by feeding them food laden with heavy metals.) Since Primor was traveling with a vial of lyophilyzed extract of pardaxin, we planned and promptly executed some experiments with test organisms.

The Atlantic sharpnose sharks, ranging in weight from 147 to 156 g, would swim about in small aquaria for considerable lengths of time. The addition of small quantities of the *Pardachirus* extract (20 g/mL seawater) caused them to behave abnormally and, after one hour, was lethal. Naturally, a substance that causes death in an hour would not be considered a "repellent." However, the behavioral response to the extract was instantaneous: The sharks became excited, swam about rapidly, and attempted to leap out of the water. There was a concomitant secretion of mucus-like material from their gills and several glandular pores associated with the lateral line system.

After the initial excited activity, the sharks settled down and exhibited a number of responses. These lasted from minutes to tens of minutes and included irregular opening and closing of the gill slits, increased mucus secretion, and finally disequilibrium and death. Similar but more rapid effects on behavior, followed by death, occurred when we placed small marine teleosts (pinfish, *Lagodon rhomboides*) in corresponding dispersions of pardaxin.

We now know a number of things about pardaxin and its pharmacological effects. Briefly, it is an acidic protein, with a rather high molecular weight (about 16,000, see Primor et al., 1978). Studies with teleosts reveal the deleterious effect it has on membranes and on ion transport processes. In particular, it is an extremely potent inhibitor of Na^+-K^+-ATPase (see Primor et al., 1980). On a molar basis, pardaxin is about 1000 times more potent than the low-molecular-weight compound, ouabain (see Figure 5), which is a natural product obtained from the plant species *Strophanthus gratus*. Disruption of ion regulation by ouabain is probably very different at the molecular level from that brought about by pardaxin.

Ion regulation is an important aspect of adaptation to the marine environment. Although all organisms have to maintain their cellular interiors with the proper balance of salts, marine organisms are especially sensitive to this problem because of the great disparity between the salt content of the seawater in which they live and that of their intracellular environment. Sharks and bony fish differ from each other with respect to how they handle the ionic problems of living in seawater. In both groups, however, active ATPases are critical to survival.

ATPases can act as molecular ion pumps. These transmembrane proteins are extremely important multisubunit enzyme complexes that maintain the osmotic balance of the interior of cells both in a qualitative and quantitative sense. Two kinds of ATPases involved in ion transport are shown in Figure 6: Na^+-K^+-ATPase, which helps to regulate ionic balance in marine organisms, and Ca^{++}-ATPase, which is associated with plasma membranes, the biochemical machinery of the sarcoplastic reticulum, and muscle contraction.

The sensing system of the ampullae of sharks can detect changes in the salinity of the external seawater (see Chapeville and Haenni, 1980; Hodgson and Mathewson, 1978; Murray, 1965). The molecular system for ion regulation may be associated with a salt-sensitive, but separate, detection system. Alternatively, the molecular pumping mechanism itself may be involved in a salt-sensitive sensing system. In any event, because of the importance of maintenance of a proper salt balance, organisms might have "learned" to stay away from chemical environments that affect their ion regulation pumps. We are investigating the possibility that a class of chemicals that adversely affect ion regulation might be effective as shark repellents. ATPase inhibition may not be an uncommon defense mechanism. The bufadienolide toxins found in toad skins are potent inhibitors of Na^+-K^+ -ATPase (Flier et al., 1980); various mercurials (Berg and Miles, 1979) and vanadate (Wierichs et al., 1980) inhibit ATPase; and penicillic acid, a cardioactive mycotoxin produced by various penicillium molds, is also a potent and active inhibitor of Na^+-K^+-ATPase (Phillips et al., 1980).

<u>Fireflies and Sharks: Where Is the Connection?</u>

Based partly on the synergistic interactions of the work done in our laboratory with flatfish secretions, the same creative research manager at the Office of Naval Research suggested that there might be some potentially fruitful interaction between people working with fireflies and marine scientists. Accordingly, we obtained a number of different preparations of freeze-dried fireflies from Steven Antonik, the supplier of firefly lanterns to various companies that used them as a basis for an extremely sensitive assay for ATP. An accumulation of anecdotal stories concerning the lack of natural predators for fireflies, plus a newly acquired knowledge that a manageable small shark species was available in Beaufort, led to the following experiments.

In tanks identical to those used in the flatfish experiments, we tested dispersions of lyophilized fireflies and firefly extracts on the Atlantic sharpnose shark (Rhizoprinodon terraenovae), the smooth dogfish (Mustelus canis), the pinfish (Lagodon rhomboides), and killifish (Fundulus heteroclitus). Toxic and behavioral effects occurred with all test organisms. Dispersed fireflies at a concentration of four fireflies per liter produced rapid behavior responses in the elasmobranchs. These responses were instantaneous, in the same sense as in the experiments with pardaxin, but after several minutes of exposure the sharks became unable to swim and unresponsive to external stimuli. Transferring them at this point to tanks devoid of the firefly material led to eventual recovery in all cases. Death did not result from exposure periods as long as an hour.

Experiments done with the teleosts were quite similar in terms of behavioral response. Short-term exposure revealed a semiparalytic but nonlethal nature of the firefly preparation. Death occurred after long-term exposure (8-24 hours) at levels of 0.1 g per liter of unpurified firefly extract in experiments with killifish. These effects may be due to firefly steroids that are chemically related to the bufadienolides found in toad skin (Eisner et al., 1978; Flier et al., 1980; Meinwald et al., 1979).

With respect to putative repellency activity of the firefly material, some tests were performed with killifish, and similar tests are planned for the elasmobranchs. A preference tank approximately 1.2 meters in length and 0.1 meters in diameter was constructed from white polyvinyl chloride tubing (see Figure 7). The pipe was sliced in half lengthwise, producing a trough containing about 4 liters of seawater. Fittings for tubing leading to a syringe pump were sealed in place at each end of the trough. The syringe pump was adjusted to yield measured inputs of material. The whole unit drained through small holes at the high-water line at the middle of the tank. Such a system produces a chemical gradient controlled by the various combined flow rates.

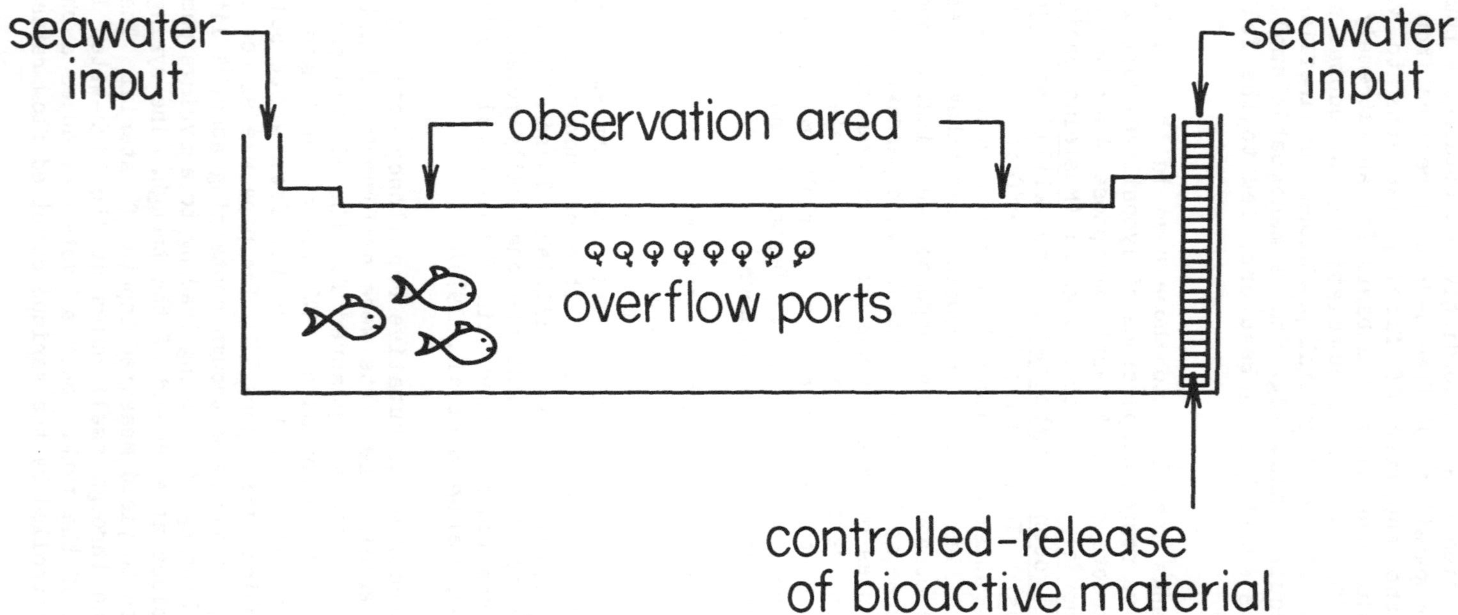

Figure 7. Schematic representation of preference tank used in testing the behavioral response of killifish to firefly extracts.

We filled the tank with 4 liters of clean seawater and set the flow rates of injected seawater at 1 mL per minute. One killifish was placed in the middle of the tank and observed for 15 minutes. Then, we introduced an ethanol-lipid extract from 20 g of fireflies at a rate of 1 mL per minute at one end of the preference tank, introduced pure ethanol at the other end, and noted the behavior of the fish for the following 15 minutes. Finally, we washed the tank clean with soap and water, followed by numerous rinsings, rotated it 180 degrees, and refilled it for the next trial. These preference tests clearly indicated that killifish preferred not to be in areas containing firefly extracts.

A Biochemical Approach to Shark Repellent Delivery Systems

Two major aspects are now clear in consideration of a biochemical approach to the development of an effective shark repellent. First, we must determine what chemical recognition system should be the target of the repellent. The concept that a class of chemicals can act as repellents by interference with ion regulation is a testable hypothesis. The concept of using inhibitory or repellent analogues to molecules that normally signal the presence of food also suggests a direction for further research. Provided a repellent chemical can be found, the next stage of the development would involve the delivery system. Should repellent chemicals be continuously dispersed or activated only when the individual in the water has cause for concern?

Delivery systems for shark repellents may take advantage of two relatively new technological advances. The first is "controlled release"--a continuous release of an active compound into the seawater (see Figure 8). Such a system might be activated by peeling off an impervious sheet, which would then allow for a controlled release of the repellent substance into the seawater. The fact that low concentrations of specific substances can be released in a continuous way by this approach is its most attractive feature. Such a mode of delivery would, of course, require that the repellent be active at very low concentrations.

We have conducted some experiments to test the possibility of slow release, utilizing killifish and firefly extracts. The firefly extracts were incorporated in a slow-release state in a hydrophilic polyurethane matrix. Activity of the firefly toxin was still expressed in long-term tests and in preference tank experiments. Although similar tests are planned with sharks, the controlled release experiments, to date, have been done only with teleosts.

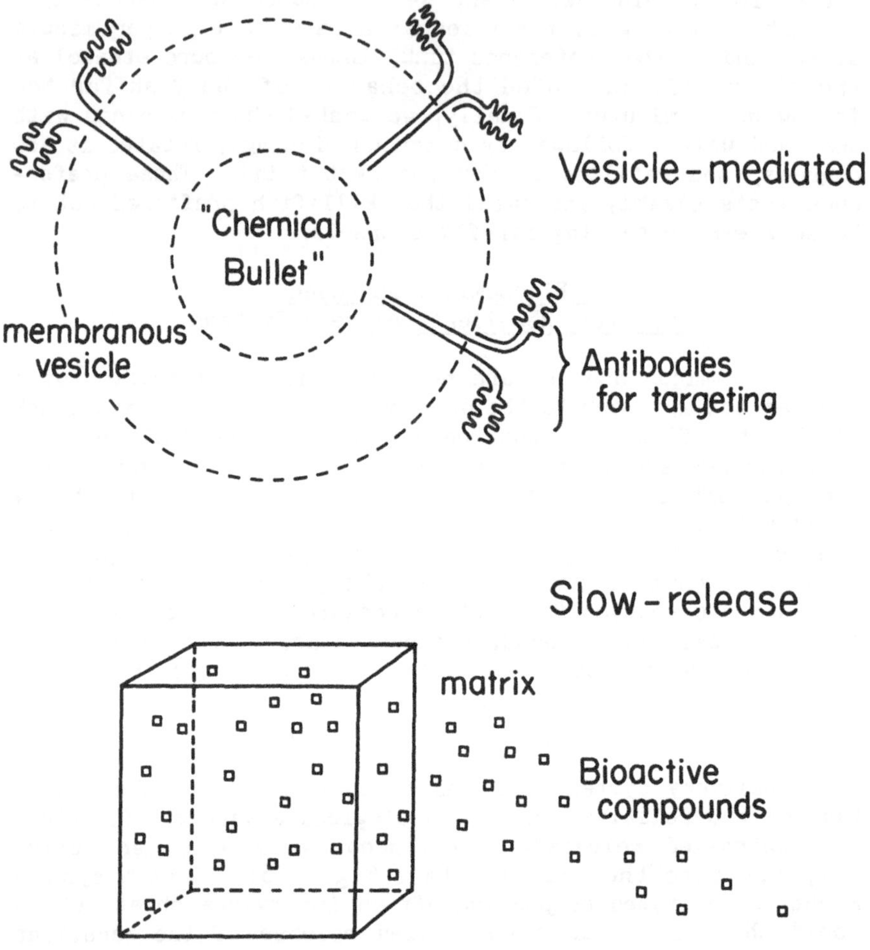

Figure 8. Possible systems for delivery of new-generation shark repellents. In slow-release systems, the active compounds are embedded in a matrix in such a way that they diffuse out at a controlled rate when placed in contact with seawater. In vesicle-mediated delivery, the active compounds are packaged in phospholipid vesicles, which can be made so they readily fuse with biological membranes and deliver their contents within the cell with which they have fused. Great specificity can be given to the vesicles by incorporation of specific antibodies.

Another delivery system might involve a new approach to drug delivery (see Gregoriadis, 1977; Pagano and Weinstein, 1978), called "vesicle-mediated transport," which involves packaging biologically active compounds in cell-like vesicles (see Figure 8). The vesicles can be made very specific in terms of cell types with which they would like to fuse. If a target receptor can be identified in sharks, we may be able to synthesize the antibodies specific for the targets that could help deliver the vesicle-mediated compound. Antibodies specific for membrane components of specific chemical sensing organs are potential candidates for testing. The wide-scale application of this new technology is made possible by work with recombinant DNA, whereby specific proteins can be synthesized in very large quantities.

Extrapolation

One problem inherent in this sort of research involves the extrapolation of laboratory studies to field or real-life conditions. Nevertheless, we must proceed and continually integrate all the experimental approaches. Extrapolation problems exist, of course, when considering bony fish versus shark behavior, small shark versus large shark behavior, and so on. One of the exciting aspects of this symposium is that it brings together scientists from a wide variety of disciplines and allows them to pool their respective knowledge, with the hope that there will be a synergism that leads to a solution to the problem at hand, including that of extrapolation.

We hope that the biochemical approach presented here will lead to extensive discussion and critical probing of premises and procedures in shark repellent research. Many of us, as a result of this symposium, will return to our home laboratories with fresh enthusiasm for a new generation of experiments, a refinement of procedures, and the eventual development of a system that truly operates as we would all envision: a simple fail-safe unit that allows humans to be in the sea, whether by choice or by accident, without one major fear of such seaborne existence--predation by sharks.

References Cited

Arnone, A. 1972. X-ray diffraction study of binding of 2,3-diphosphoglycerate to human deoxyhaemoglobin. Nature 237:146-149.

Berg, G.G., and E.F. Miles. 1979. Mechanisms of inhibition of active transport ATPase by mercurials. Chem. Biol. Interactions 27:199-219.

Chapeville, F., and A.L. Haenni, eds. 1980. Chemical Recognition in Biology. Springer-Verlag, Berlin.

Clark, E. 1974. The Red Sea's sharkproof fish. Natl. Geogr. 145(5):718-787.

Clark, E., and S. Chao. 1973. A toxic secretion from the Red Sea flatfish, Pardachirus marmoratus (Lacepede). Bull. Sea Fish. Res. Stn. (Haifa) 60:53-56.

Clark, E., and A. George. 1979. Toxic soles, Pardachirus marmoratus from the Red Sea and P. pavoninus from Japan, with notes on other species. Environ. Biol. Fishes 4(2):103-123.

Eisner, T., D.F. Wiemer, L.W. Haynes, and J. Meinwald. 1978. Lucibufagins: defensive steroids from the fireflies Photinus ignitus and P. marginellus (Coleoptera: Lampyridae). Proc. Natl. Acad. Sci. USA 75:905-908.

Flier, J., M.W. Edwards, J.W. Daly, and C.W. Myers. 1980. Widespread occurrence in frogs and toads of skin compounds interacting with the ouabain site of Na^+, K^+-ATPase. Science 208:503-505.

Gregoriadis, G. 1977. Targeting of drugs. Nature 265:407-411.

Hodgson, E.S., and R.F. Mathewson, eds. 1978. Sensory Biology of Sharks, Skates, and Rays. Department of the Navy, Office of Naval Research, Arlington, Va.

Meinwald, J., D.F. Wiemer, and T. Eisner. 1979. Lucibufagins, Part 2. Esters of 12 oxo-2-beta 5-beta 11-alpha-trihydroxy bufalin, the major defensive steroids of the firefly Photinus pyralis (Coleoptera: Lampyridae). JACS 101:3055-3060.

Murray, R.W. 1965. Sensory receptors. Cold Spring Harbor Symp. Quant. Biol. 30:233-243.

Pagano, R.E., and J.N. Weinstein. 1978. Interactions of liposomes with mammalian cells. Annu. Rev. Biophysics Bioeng. 7:435-468.

Parness, J., and E. Zlotkin. 1976. Action of the toxic secretion of the flatfish Pardachirus marmoratus (Soleidae). Toxicon 14:85-91.

Phillips, R.D., P.K. Chan, and A.W. Hayes. 1980. Biochem.

Pharmacol. 29:19-26.

Primor, N., and E. Zlotkin. 1975. On the ichthyotoxic and hemolytic action of the skin secretion of the flatfish Pardachirus marmoratus (Soleidae). Toxicon 13:227-231.

Primor, N., J. Parness, and E. Zlotkin. 1978. Pardaxin: the toxic factor from the skin secretion of the flatfish Pardachirus marmoratus (Soleidae). Pages 539-547 in P. Rosenberg, ed. Toxins: Animal, Plants, and Microbial. Pergamon Press, Oxford.

Primor, N., I. Sabnay, V. Lavie, and E. Zlotkin. 1980. Toxicity to fish, effect on gill ATPase and gill ultrastructural changes induced by Pardachirus secretion and its derived toxin pardaxin. 211:33-43.

Wierichs, R., A. Hagenmeyer, and H. Bader. 1980. Influence of calcium and magnesium on the vanadate inhibition of the calcium -ATPase from pig heart sarcoplasmic reticulum. Biochem. Biophys. Res. Commun. 92:1124-1129.

6. Shark Repellent Effect of the Red Sea Moses Sole

Soles of the genus Pardachirus (Family Soleidae) can be distinguished from all other flatfishes (Order Pleuronectiformes) by the presence of a series of pores at the bases of the dorsal and anal fins. Clark and Chao (1973) demonstrated that the milky fluid from these pores in the Red Sea species, the "Moses sole"--Pardachirus marmoratus (Lacepede)--was toxic and lethal to teleost fishes. This led to studies on the shark repellent effects of the toxin (Clark, 1974), to comparisons of the Moses sole with other species in the genus (Clark and George, 1976, 1979), and to a series of chemical and pharmacological studies on the P. marmoratus toxin by Primor and other investigators (see reviews in Primor et al., 1978, and Clark and George, 1979).

This report is on shark repellent tests made at the Heinz Steinitz Marine Biology Laboratory in Elat and field tests made at the southern tip of the Sinai peninsula at Ras Muhammed. Three types of tests were conducted: (a) Moses sole specimens were presented to captive sharks; (b) nontoxic teleost fishes bathed in dilutions of the P. marmoratus secretion were presented to captive sharks; and (c) a line of ten teleost fishes, including the Moses sole, was presented to sharks in the open sea.

Material and Methods

In 1972 and 1973 tests on captive sharks were made at the Heinz Steinitz Marine Biology Laboratory of Hebrew University at Elat, Israel. Two female reef whitetip sharks, Triaenodon obesus (Rüppell), caught locally in 1970, were kept in an outdoor, shaded, cement holding tank (3 x 3 meters, water depth 96 cm) with continuously running seawater. The seawater was turned off when observations on sharks were being recorded. The sharks were approximately 110 and 125 cm in length. Also present in the tank were a discfish or

Figure 1. The Moses sole is a relatively slow swimmer. Its more than 200 poison glands occupy some of the space of swimming muscles that operate the movements of the dorsal and anal fins. Here a specimen caught by hand is released before being grabbed again.

remora, Echeneis naucratus (about 40 cm long) and a hawksbill turtle (carapace length about 35 cm). The turtle was removed during tests after 1972.

These animals were regularly fed small or cut fish several times a week. For test periods feeding was discontinued several days in advance and during tests, except for "control fish." Control fish consisted of small whole or cut pieces (10-18 cm) of teleost fishes (fresh or defrosted) of the genera Siganus, Upeneus, Lethrinus, Gerres, Sparus, and Bothus. A control fish was presented before or at the start of each test and, if it was not eaten by the sharks within 3 minutes, the test was postponed. Fish offered to the sharks were tossed in near the center of the tank, or suspended there on a thread within 30 cm of the water surface. P. marmoratus (males and females 12.3 to 20.5 cm standard length) caught near Elat (Figure 1) was always suspended underwater on a string that had been threaded through the gill opening and out the mouth with a blunt needle. This string in turn was tied onto the center of a second string across the top of the tank. Control fish were strung the same way except by only one long string held by the experimenter. Underwater photographs were taken with a strobe light and Nikon camera in an Ocean Eye housing on tripods, with a remote-control cable release.

The Moses sole (see Figure 1) were offered to the sharks in several ways: alive and therefore wriggling, at least intermittently, setting up the low frequency vibrations of a struggling fish; freshly dead, intact or sometimes after the toxic secretion was partly "milked" from the pores; or frozen then freshly defrosted without removing any toxic secretion. The amount of toxic solution that can be pressed out of a large Moses sole is about 2 mL. "Milking" the sole was done by blotting the skin dry with a paper towel, pressing on the glands (see Figure 2), and drawing the secretion into a pipette (Clark, 1974; Clark and George, 1979). The secretion was diluted with seawater (see Figure 3).

In the shark repellent experiments where nontoxic fishes (bait) were bathed in dilutions of the toxic secretion of P. marmoratus, we followed a method similar to that of Gilbert and Springer (1963, p.482), who tested chemicals on sharks in enclosures in the Bahamas. We placed a 1.2 cm diameter teflon tube, carrying a continuous supply of seawater, directly over the bait. When an attack on the bait was imminent (head of shark approached within 20 cm), a known dilution of the toxic secretion was injected with a hypodermic syringe into the upper end of the teflon tube. The secretion was then carried along by the seawater (flowing

Figure 2. The underside of a Moses sole seen through a plastic bag with the anemone fish, Amphiprion bicinctus. The latter died less than a minute after this close contact with the Moses sole. Pores from the sole's poison glands open at the base of each dorsal (lower) and anal (upper left) fin rays. White poison glands can be seen through the translucent skin of this left, unpigmented side of the sole (pale "ampules" under the belly region of the anemone fish).

Figure 3. The toxic secretion (approx. 2 mL) pressed out from one Moses sole retains its milky appearance when diluted with seawater.

TEST NO.	SOLE SEX	SOLE mmSL	DATE	TIME START	NO. OF ATTACKS*
1	♂	122	2-7-72	1300	4
2	♀	146	25-12-72	1308	17
3	♀	146	25-12-72	1415	58
4	♀	185	26-12-72	1215	6
5	♀	178	6-1-73	1617	6
6	♀	166	16-7-73	1820	3
7	♂	138	17-7-73	1440	2
8	♀	174	10-8-73	1840	47
9	♀	165	11-8-73	1145	12
10	♀	205	12-8-73	2030	48
11	♀	165	13-8-73	2051	24

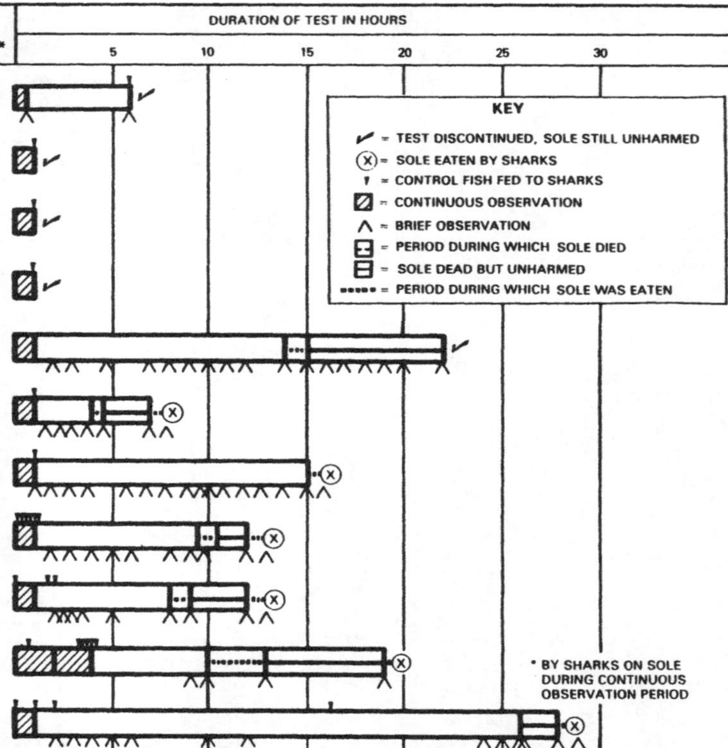

Figure 4. Shark repellent effect of live Moses sole *Pardachirus marmoratus*) on two captive sharks (*Triaenodon obesus*). (*Attack = fast swim towards sole with head of shark approaching within 10 cm of sole.)

continuously in the tube) onto the bait with no change in turbulence. We terminated each experiment (after several attack approaches were aborted by toxic injections) by allowing a shark to take the bait without bathing it in toxic secretion. We observed the activity patterns of sharks in the cement tank, before and after presenting the bait, by dividing the water surface into 9 square meters (using four strings tied across the top, tic-tac-toe fashion) and recording how many square meters a shark passed through in a unit of time.

In our field tests at Ras Muhammed, we hung a main line near the surface between coral projections in water more than 50 meters deep. The main line had ten side lines, each 20 cm long, spaced 80 cm apart. A bait fish was tied to the end of each side line. Bait fish were usually defrosted Caesio species but we also used other nontoxic teleost fishes of the families Mugilidae, Mullidae, Scaridae, and Labridae.

Results

Figure 4 summarizes the results of eleven tests on the shark repellent effect of live Moses sole presented to two captive sharks. Tests 1 to 5 were terminated to examine the condition of the sole. In each case the sole did not appear harmed by the sharks; not one scratch was evident, even at the end of test 5 at which time the sole had been dead at least 7 of the last 22 hours of its exposure to the sharks. In tests 6 to 11 the sole was left in the tank until one of the sharks ate it. In test 6 we removed the sole briefly from the tank when it was first found dead; some milky secretion could still be squeezed out, showing no depletion of toxic secretion during 4 hours of exposure to attacking sharks.

During the first half to four hours of each test we continuously recorded the sequence and duration of behavior patterns of the fishes. "Hunting turns" (Gilbert and Springer, 1963)--the increased swimming speed and tight turns of sharks near the bait (see Figure 5)--sometimes started within a few seconds after we introduced the bait. Sharks approached within 10 cm of the bait fish or mouthed it an average of 17.7 times per hour. Of 116 such attack approaches recorded on soles within the first hour, a shark drew near to the sole 85 times with its mouth closed (Figure 6); 21 times with its mouth open, but not taking the sole in its jaws (Figure 7); and 10 times taking the sole in its open jaws (Figure 8). In 31 of the total 216 attack approaches, the shark showed a strong repellent reaction involving one or a combination of the following: violent shaking of the head

Figure 5. Two female reef whitetip sharks (Triaenodon obesus) approach a live Moses sole, struggling and tied on a line in test 10 (see Figure 4). Larger shark (on right) has a remora (Echeneis naucratus) attached to its belly.

Figure 6. Reef whitetip shark turns directly toward a struggling Moses sole, before turning aside.

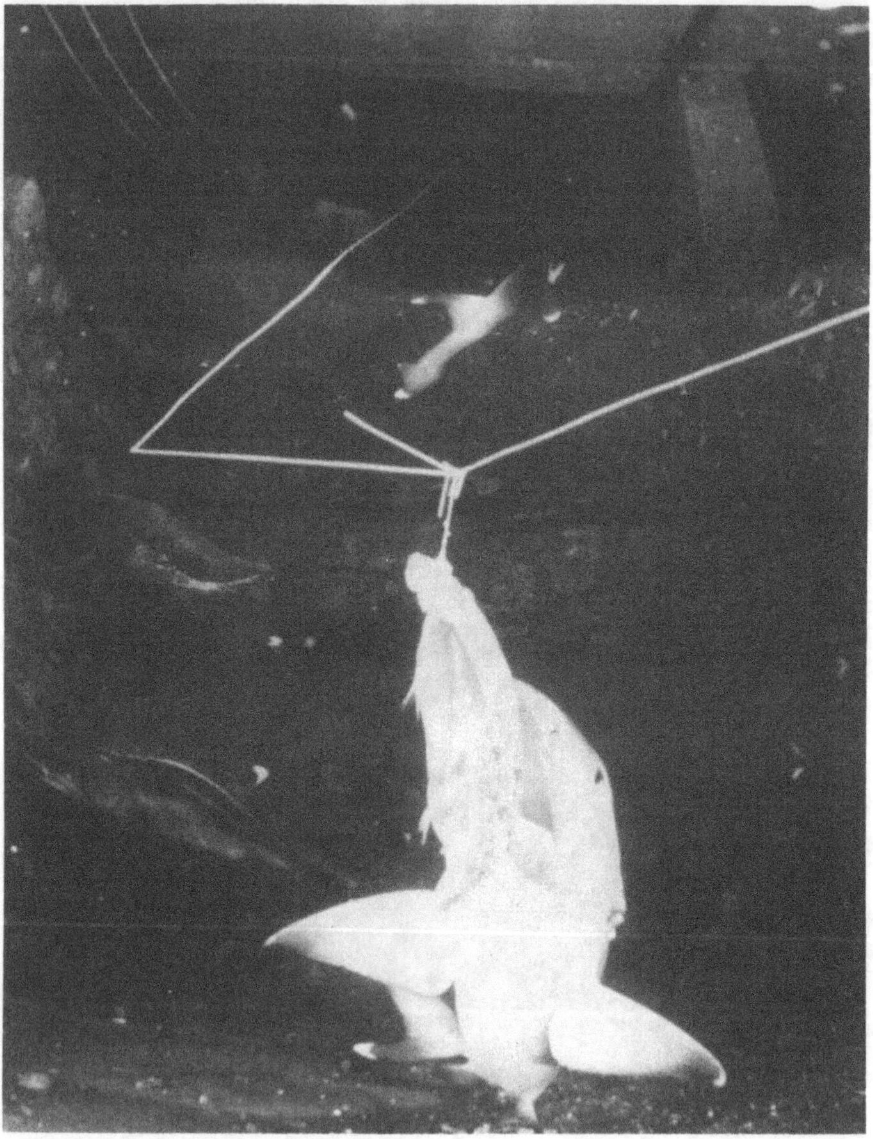

Figure 7. A reef whitetip shark approaches a Moses sole with its mouth open.

Figure 8. A reef whitetip shark with its mouth around a Moses sole. (The shark's thick, white nictitating membrane has closed over its eye.) The shark is making an abrupt stop from a fast swim, rotating its pectoral fins nearly 90°; the remora hanging on its belly has been swung around 180°. The shark did not complete this bite, but jerked away violently, hitting itself against the walls of the cement tank with its mouth still open. The sole was not harmed or scratched by the shark's teeth.

from side to side; jerking away and dashing around the tank, sometimes banging into the walls; holding its mouth open for several seconds to more than a minute; swimming or curling on the bottom of the tank, belly-up, sometimes with its mouth held open, in one case for more than three minutes. In only two instances did we see the milky secretion released into the water by the Moses sole, both times after a series of attack approaches by the sharks.

Control fish were usually eaten by the sharks in less than 1 minute (2 sec. minimum, 2 min. 25 sec. maximum, 50 sec. $\bar{x}$, N = 33). Moses sole, presented alive, were not eaten for at least 7 to 28 hr. ($\bar{x}$ 14 hr., N = 6) and, in most cases, were consumed only after they had been dead for hours.

In similar tests we offered the sharks dead, or nearly dead, Moses sole after we had already squeezed out much of their toxic secretion (for other experiments not reported here). The sharks ate them usually in less than 2 hr. (40 sec. to 360 min., $\bar{x}$ 92 min., N = 8). In the two shortest tests (40 sec. and 9 min.) the toxic secretion was coagulated in the glands and was not fluid and milky when pressed out, as it was in the sole specimens used in the other 6 tests (which ran 25 min. to 3 hr.).

In tests we presented the sharks with a freshly dead Moses sole that had not been milked of its toxic secretion but had the mucus on its body removed with ethyl alcohol. In each case the shark ate the sole within 1 minute.

In another series of tests we tied dead goatfish (Upeneus) progressively closer to a live Moses sole and obtained the following results: A goatfish tied 100 cm from the sole was eaten in 14 sec.; one at 30 cm from sole in 1 min. 26 sec.; and two at 10 cm in 3 min. 40 sec. and 4 min. 35 sec.. Goatfish tied less than 10 cm were not consumed by the sharks during four 1-hour tests. However, two goatfish soaked in 50% and 4% solutions of toxic secretion, then immediately put into the tank with the sharks, were eaten in 26 sec. and 1 min. 20 sec., respectively.

The experiments that presented captive sharks with nontoxic bait fish bathed in a stream of running seawater (control tests) from a teflon tube produced notable results. In repellent tests, we injected a 1% solution of P. marmoratus secretion into the teflon tube, which therefore became further diluted in the running seawater and in the tank water surrounding the bait fish hanging under the lower opening of the tube. In 11 control tests (before and after tests with toxic solutions) the sharks ate the bait fish in about 1 minute (3 sec. to 2 min. 40 sec., $\bar{x}$ 1 min. 06 sec.; N = 11). In 5 tests with diluted P. marmoratus secretion running over the bait fish, the sharks closely approached the bait fish 60 times during 38 min. 10 sec. but never ate them.

Field tests at Ras Muhammed, with a line of 10 bait fish set overnight (Figure 9), attracted four species of sharks (Carcharhinus albimarginatus, C. wheeleri, C. limbatus, and Triaenodon obesus). Preliminary tests run during the day attracted too many small teleost fishes (e.g., surgeon fish), which nibbled away most of the nontoxic bait fish before the sharks arrived. Thus, we set the 11 main test lines of 10 fish each in the late afternoon (between 1500 and 1800 hr.) 4 during winter 1972 and 7 during summer 1973. The difficult field conditions for the winter experiments (e.g., only three observers, plus weather and army security problems) inhibited our checking the lines at regular intervals, but we found all the fish removed after 24, 15, 17, and 16 hr. (both nontoxic fish and Moses sole).

In summer 1973, six of us camped near the lines and took turns checking them frequently. In 1 test the entire line disappeared shortly after it was set, probably taken by a shark. In the other 6 tests we monitored the set lines for periods of 12 to 19 hr. ($\bar{x}$ 15) and used 25 Moses sole (20 dead and 5 alive) and 35 nontoxic dead bait fish. Sharks and one large moray eel usually started eating the bait fish off the lines (starting with the nontoxic specimens) within the first few hr. after the line was set. The predators disposed of the 25 Moses sole as follows: 16 unconsumed; 2 eaten after 10 hr.; 5 after 5 hr.; and 2 after 3 hr. None was eaten in less than 3 hr. In two tests we tied a grouper on a side line with a Moses sole, and in both cases neither fish was taken after 15 and 16.5 hr.

Conclusions

From all these experiments, we drew the following conclusions:

- Live Moses sole (Pardachirus marmoratus) presented to sharks in captivity can repel the sharks 7 to 28 hr. even though the sole die during the last few hr. Dead Moses sole with part of their toxic secretions squeezed out repel sharks for only about 2 hr.

- Bait fish bathed in a dilute running solution of P. marmoratus toxic secretion are 100% effective in repelling sharks.

Figure 9. A shark line, set near the surface at Ras Muhammed, is baited with one Moses sole and nine nontoxic teleost fishes.

• Live and dead specimens of Moses sole tied to a set line with nontoxic bait fish repelled at least four species of sharks in the open sea for periods of 3 to 16.5 hr., usually more than 10 hr.

• The repellent effect of Moses sole, which works on other predators and is toxic to fishes and invertebrates, is due to a toxic secretion in small amounts retained in the mucus covering of the sole. When this mucus is removed, the sole no longer can repel predators.

• An area approximately 10 cm surrounding a Moses sole repels sharks; they will not take nontoxic bait fish within this area. In fact, sharks are deterred from entering an area up to 30 cm around the Moses sole.

• The mechanism of the Moses sole, effective in repelling sharks for hr. with only minute amounts of toxin retained in its mucous coating rather than dispersed directly into seawater, makes the practicability for developing chemical shark repellents more promising than has been suggested by Baldridge (1976).

Acknowledgments

This study was supported by the National Geographic Society and the U.S. Office of Naval Research. I am grateful to many who aided this work: F.D. Por and the staff of the Heinz Steinitz Marine Biology Laboratory of Hebrew University in Elat, Israel, extended laboratory facilities and cooperation; David Fridman was most helpful in collecting specimens; and Abshalom, Avi Barnes, Willard Cook, Nikolas Konstantinou, and Gail Weinmann helped in laboratory and field experiments. The photographs taken by David Doubilet documented these studies and allowed a more detailed analyses of shark reactions. Taz Rufty prepared Figure 4.

References Cited

Baldridge, H.D., Jr. 1976. A reminder of the impracticability of chemical shark repellents. Page 18 in W. Seaman, Jr., ed. Sharks and Man: A Perspective. Florida Sea Grant Report No. 10, Gainesville, Fla.

Clark, E. 1974. The Red Sea's sharkproof fish. Natl. Geogr. 145(5):718-727.

Clark, E., and S. Chao. 1973. A toxic secretion from the Red Sea flatfish, Pardachirus marmoratus (Lacepede). Bull. Sea Fish. Res. Stn. (Haifa) 60:53-56.

Clark, E., and A. George. 1976. A comparison of the toxic soles Pardachirus marmoratus of the Red Sea and P. pavoninus from southern Japan. Rev. Trav. Inst. Marit. 40:545-546 (abstract).

_____. 1979. Toxic soles, Pardachirus marmoratus from the Red Sea and P. pavoninus from Japan, with notes on other species. Environ. Biol. Fishes 4(2):103-123.

Gilbert, P., and S. Springer. 1963. Testing shark repellents. Pages 477-494 in P.W. Gilbert, ed. Sharks and Survival. D.C. Heath and Co., Boston.

Primor, N., J. Parness, and E. Zlotkin. 1978. Pardaxin: the toxic factor from the skin secretion of the flatfish Pardachirus marmoratus (Soleidae). Pages 539-547 in P. Rosenberg, ed. Toxins: Animal, Plants and Microbial. Pergamon Press, Oxford.

Naftali Primor, Celia Bonaventura,
Joseph Bonaventura

7. Effect of Moses Sole Secretion and Its Active Factor, Pardaxin, on Elasmobranchs

Shark attacks have always presented a problem to man's marine activities. Our increased interest in and need for exploration and use of the seas have spurred the search for a solution. One possibility lies in finding a chemical compound to which sharks have an aversion.

At present, chemicals are not used to prevent shark attacks, although investigators have tested the shark-repellent properties of many man-made and naturally occurring substances (Gilbert and Gilbert, 1973). The difficulty with this kind of solution lies primarily in the large quantity of chemicals needed to compensate for their dilution in water (Baldridge, 1969a, 1969b). Recently, however, researchers have developed methods to trap an active chemical in an insoluble material. With this technology we can immobilize a potent shark repellent, which, when released, would give an effective local concentration.

The search for shark-repellent chemicals was recently stimulated by the finding that the secretion of the Red Sea flatfish *Pardachirus marmoratus* can repel a predatory shark (Clark, 1974). In this paper we review the action of that secretion and its active fraction--pardaxin--on elasmobranchs under laboratory conditions. We used two elasmobranch species--the Atlantic sharpnose (*Rhizoprionodon terraenovae*) and the dogfish shark (*Squalus acanthias*)--in our experiments.

Response of the Atlantic Sharpnose to *P. marmoratus* Secretion

The Atlantic sharpnose shark is a common species that appears in the waters in the vicinity of Beaufort, North

Carolina. The average adult weighs 150 g. This shark is readily caught by trawl and can be maintained for long periods of time in holding tanks in the laboratory. In fact, this hardy species can be extensively manipulated without serious adverse effects. That is, drastic procedures, such as removing them from holding tanks, taking blood samples, returning them to the tanks, and so on, do not significantly disrupt the animals. These properties make the Atlantic sharpnose particularly well-suited for laboratory studies, especially in cases like ours, in which the compound is available in limited quantities for testing with fish that require large amounts of seawater.

To test the effect of Pardachirus marmoratus secretion on these sharks, we used cylindrical tanks 50 cm in diameter. Control experiments determined that, with 6 liters of unaerated and unstirred seawater in the tank, one Atlantic sharpnose shark could live comfortably for at least one hour. Observations made for several hours after subsequent transfer to holding tanks revealed no adverse effects from the hour spent in the test tank. When we then placed an Atlantic sharpnose into a testing tank containing seawater and 10 μg/mL of P. marmoratus secretion, there was an immediate brief burst of hyperactivity. Thereafter, the sharks exhibited labored ventilation and reduced sensitivity to external stimuli. Concomitantly, a mucus-like secretion from gills and epidermis appeared. After one hour we rinsed the shark with uncontaminated seawater and transferred it to a holding tank with flowing seawater. Although the mucoid secretion persisted for several hours, after another hour of "recovery" in the holding tank, the shark appeared to behave relatively normally.

Increasing the concentration of P. marmoratus secretion to 20 μg/mL produced approximately the same results, except that (a) the mucoid secretions were more pronounced; (b) 20 minutes of exposure caused irregular opening and closing of opercular slits; (c) 30 minutes of exposure caused definite disequilibrium; and (d) 45 and 50 minutes of exposure resulted in death.

In another test we took blood samples from one shark at 10-minute intervals during its exposure to the secretion. The mean sodium and potassium plasma concentrations were 228.6 and 4.8 mM, respectively, as measured at 10, 20, 30, 40, and 50 min. Since the control values were 249.3 and 5.7 mM, the test indicated no apparent change in the concentration of the shark's electrolytes.

The Responses of the Dogfish Shark to Pardaxin

To determine the target organ for pardaxin's action, we placed an adult dogfish (4.0-5.0 kg) in a large apparatus fitted with a rubber diaphragm inserted close behind the posterior gill cleft (see Figure 1). Pardaxin (25 μg/mL) administered to the head region instantly caused the shark to struggle and to open its mouth as if to gasp. Although the opercular rate initially dropped from 40-42/min to 8-11/min, it recovered substantially, despite the continuing presence of pardaxin. When we introduced pardaxin in the seawater bathing the rear part of the dogfish, there were no behavior responses (Primor et al., 1980a).

Because the head region of the dogfish was the target of pardaxin's action, and because its gills are relatively impermeable to urea (Boylan, 1963), we decided to study whether pardaxin could induce changes in gill permeability to urea and sodium. No measurable radioactivity was detected in the diaphragm chambers during stoppage in seawater flow in control fish injected with 200 μCi ^{14}C-urea or $^{22}Na^{+}$. When we administered pardaxin (25 μg/ML) to the head section of the apparatus, a marked leakage of ^{14}C-urea and $^{22}Na^{+}$ leaked from the plasma to the seawater and ^{14}C-urea moved from the seawater to the blood. After reaching a maximum, these leakages decreased, indicating a reversible character of pardaxin action on the gills. Further, the pardaxin concentrations of 25 μg/mL were not lethal, and the tested fish were able to recover.

On the other hand, no measurable radioactivity in seawater was observed when 40 mg of pardaxin was injected into fish previously injected with ^{14}C-urea or $^{22}Na^{+}$. In addition, pardaxin injection did not cause struggling or mouth opening or changes in the operculum rate (Primor et al., 1980a).

Pardaxin was, however, toxic to dogfish pups when administered via the blood; so we tested its effect on isolated epithelial tissue when introduced in this manner (Primor et al., 1980b). Silva et al. (1977) had demonstrated that the isolated rectal gland secretes chloride against its electrochemical gradient and shows a very low permeability to urea. Thus, we injected pardaxin into the artery of the isolated perfused rectal gland. The secretion of urea in the rectal gland fluid increased at a rate dependent on the pardaxin dosage; saturation occurred at 400 μg. The urea gradient between the perfused Ringer (350mM) and the rectal gland fluid (8-20 mM) was abolished, and the chloride gradient between the rectal gland fluid (480-500 mM) and the perfused

Figure 1. Responses of the dogfish shark to pardaxin. The 5 kg shark was positioned in an apparatus with a rubber diaphragm inserted close to the posterior gill cleft, separating the head region from the rest of the body. Pardaxin (25 μg/mL) administered to the head region caused the fish to struggle and open its mouth and diminished the opercular rate. None of these responses occurred when pardaxin was administered to the rear part of the apparatus.

Ringer (290 mM) was also eliminated. The injected pardaxin caused significant amounts of mucus to appear in the rectal gland fluid. In contrast to the effects on the gills, the increased permeability to urea in the rectal gland was not reversible; thus, pardaxin apparently caused permanent damage to the rectal gland tissue (Primor et al., 1980b).

Summary and Discussion

Since exposure to pardaxin did not affect the sodium and potassium levels in the plasma of the Atlantic sharpnose shark, we concluded that pardaxin-induced behavioral responses are not caused by a gross imbalance in the composition of the shark's electrolytes. The rapid behavioral responses may be due to pardaxin's action on the sensory cells located on the surface of the shark's gills. Imbalance of its electrolytes resulting from an increase in the gill permeability (Primor et al., 1980a) may be involved in the toxic process, but does not elicit behavioral responses directly. Raising the seawater temperature from 15°C to 25°C itself caused a substantial increase in the permeability of dogfish gills to urea (Boylan, 1963) without, however, producing the behavioral responses characteristic of pardaxin's action.

The *in vivo* gill permeability studies in the dogfish shark show that the pardaxin-induced increased permeability to urea and to sodium was reversible. The radically diminished opercular rate (Primor et al., 1980a) also recovered. However, although injected pardaxin did not elicit behavioral responses, in high concentration it was toxic (Primor et al., 1980b). Therefore, the biological significance of the *Pardachirus* secretion is primarily as a rapid repellent.

Since pardaxin did not produce any unpleasant effects on investigators who sometimes immersed their hands in 100 μg/mL solutions (Primor, personal observation), it seems to be a suitable model for an ideal shark repellent. Although it is unlikely that pardaxin could actually be used for this purpose--particularly since it is derived from a relatively rare fish--understanding its mode of action may lead to the discovery of more accessible and effective shark-repellent compounds.

Acknowledgments

We acknowledge the skillful technical assistance of Patrick Whaling and Michael Rentzepis. We thank Anne Walter for performing the ion determinations and Joan Bates for typing the manuscript. We gratefully acknowledge the Office of Naval Research (Oceanic Biology Program) for organizing and assisting in this work.

References Cited

Baldridge, H.D., Jr. 1969a. Kinetics of onset of responses by sharks to waterborne drugs. Bull. Mar. Sci. 19(4): 880-896.

____. 1969b. Analytic indication of the impracticability of incapacitating an attacking shark by exposure to waterborne drugs. Mil. Med. 134(12):1450-1453.

Boylan, J.W. 1967. Gill permeability in *Squalus acanthias*. Pages 187-195 in P.W. Gilbert, R.F. Mathewson, and D.P. Rall, eds. Sharks, Skates and Rays. Johns Hopkins Press, Baltimore.

Clark, E. 1974. The Red Sea's sharkproof fish. Natl. Geogr. 145(5):718-727.

Gilbert, P.W., and C.K. Gilbert. 1973. Sharks and shark deterrents. Underwater J. 5(2):69-79.

Primor, N., J.A. Zadunaisky, V.H. Murdaugh, and J.N. Forrest. 1980a. A study of pardaxin toxicity in the dogfish. Mt. Desert Is. Biol. Lab. (MDIBL) 20:120-122.

Primor, N., J.A. Zadunaisky, J.N. Forrest, and V.H. Murdaugh. 1980b. The effect of pardaxin on the dogfish isolated rectal gland. Mt. Desert Is. Biol. Lab. (MDIBL) 20:118-120.

Silva, P., T. Stoff, M. Field, L. Fine, J.N. Forrest, and F.H. Epstein. 1977. Mechanisms of active chloride secretion by shark rectal gland: role of Na, K-ATPase in chloride transport. Am. J. Physiol. 233:R298-R306.

8. On the Membranal Action of Pardaxin

The present short survey deals with some basic aspects and hypotheses concerning the structure/function relationships of pardaxin. Past research has shown that the toxic secretion of the Red Sea flatfish Pardachirus marmoratus (Moses sole) (a) was lethal to fishes when added to the surrounding water (Clark and Chao, 1973; Primor and Zlotkin, 1975); (b) was repellent to sharks when it contaminated their food (Clark, 1974); (c) was cytolytic, as shown on human erythrocytes (Primor and Zlotkin, 1975) and cultured chick embryo heart and mouse lymphoma cells (Zlotkin et al., 1979); (d) could induce histopathological changes in preparations such as fish gills (Primor et al., 1980) and frog neuromuscular function (Spira et al., 1976); (e) demonstrated an excitatory neurotoxic effect, as expressed in the induction of transmitter release in vertebrate neuromuscular preparations (Parness and Zlotkin, 1976; Spira et al., 1976) and a depolarizing blockage of action potentials in the cultured chick embryo heart cells (Zlotkin et al., 1979); and (f) strongly inhibited membrane-bound enzymes such as fish and crustacean gill Na and K-ATPase (Primor et al., 1980).

The vast majority of these effects were reproduced by pardaxin--the toxin derived from the P. marmoratus secretion. This concerns the ichthyotoxic, hemolytic, and excitatory action on a nerve muscle preparation (Primor et al., 1978) as well as the cytolytic, heart-cell-depolarizing (Zlotkin et al., 1979), enzyme-blocking (Primor et al., 1980), and fish-gill-histopathologic activities (Lavie, in preparation). Thus pardaxin, which is quantitatively the dominant factor in the secretion (Primor et al., 1978), in practice, is responsible for that pharmacology of the Moses sole secretion.

Some effects are related to the structural damage of tissues and cells. When applied to a frog nerve muscle preparation (Spira et al., 1976), the secretion has caused

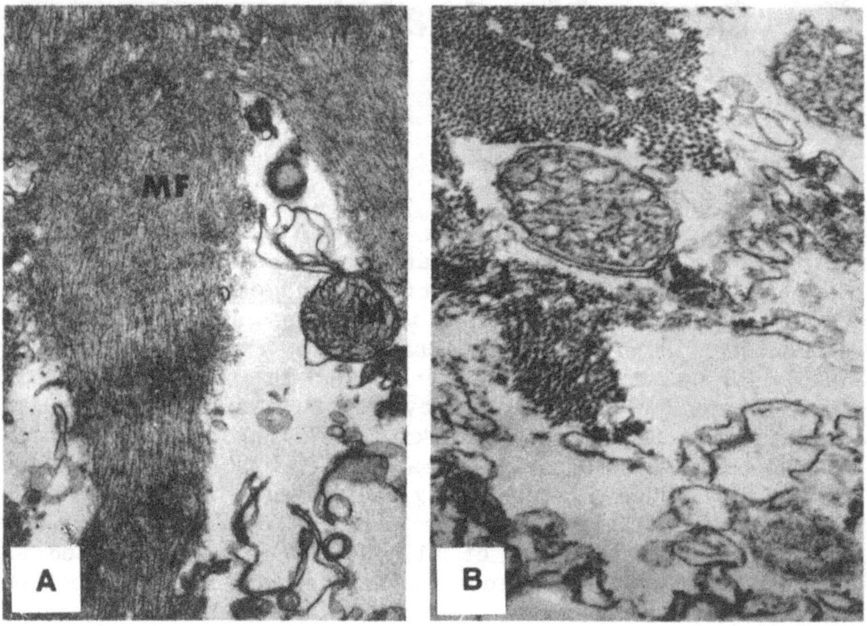

Figure 1. A. Longitudinal section of frog muscle treated by 100 μg/mL of P. marmoratus secretion for 160 min at 15°C. (x 27,000) B. Cross-section of frog muscle treated by 50 μg/mL of secretion for 80 min at 22°C. (x 36,000) Notice disarrangements of muscle filaments, spacings, and shreds of destroyed membranes and swollen mitochondria. (Reprinted by permission from Spira et al., 1976.)

(a) prolonged release of transmitter (as demonstrated by the miniature endplate potential recordings); (b) spontaneous muscle tremor and fibrillation, which were only partially blocked by D-tubocurarine, thus indicating a direct post-synaptic muscular effect; and finally (c) a blockage of the neuromuscular transmission. These effects were accompanied by a long series of morphological abnormalities such as clustering of synaptic vesicles, shrinkage of terminal cytoplasm, abnormal protrusions of Shwann cells, and damaged and swollen mitochondria. Histopathological changes, however, have occurred not only in the presynaptic region of the neuromuscular junction, but also in the muscle itself. This is indicated in Figure 1, which shows disarrangement of muscle filaments, spacings, and shreds of destroyed membranes with swollen and affected mitochondria. A damaged fish-gill epithelium is shown in Figure 2, which includes the so-called chloride and nondifferentiated cells, indicating disruption and extension of the tubular system, swelling of mitochondria, and clear damage to the nuclei of the cells. Effects of this kind were most recently demonstrated (Lavie, in preparation) in gill tissues of a euryhaline fish (Aphanias dispar) treated with pardaxin. It affected all kinds of cells present in the gill epithelium.

Figure 3 presents the cytolytic action of pardaxin to lymphoma cultured cells (Lazarovici, unpublished); swollen, vacuolated cells and fragments of disrupted cells are characteristic phenomena. In a concentration of 10-15 μg/mL during 20-25 min of application, pardaxin caused an irreversible blockage of the pulsations of cultured chick embryonic heart cells due to a gradual depolarization of the resting potential and the resulting attenuation of the action potentials. These phenomena were also accompanied by morphological-structural deformations of the cultured heart cells. These studies indicate, therefore, that the pharmacologic action of pardaxin is related to the general ability to interfere with the structural integrity of biological membranes.

The Amphipatic Character of Pardaxin

How can we explain such a diverse and versatile pharmacology of pardaxin? What is the unifying factor to the lethal, cytolytic, neurotoxic, and enzyme-blocking activities in such a wide variety of biological systems and preparations? We suggest that the effects of pardaxin are basically due to an interaction with membranal phospholipids. Such interactions may induce dose- and/or time-dependent effects ranging from a complete disintegration of the membranes (expressed in the lytic and gross-histopathological phenomena)

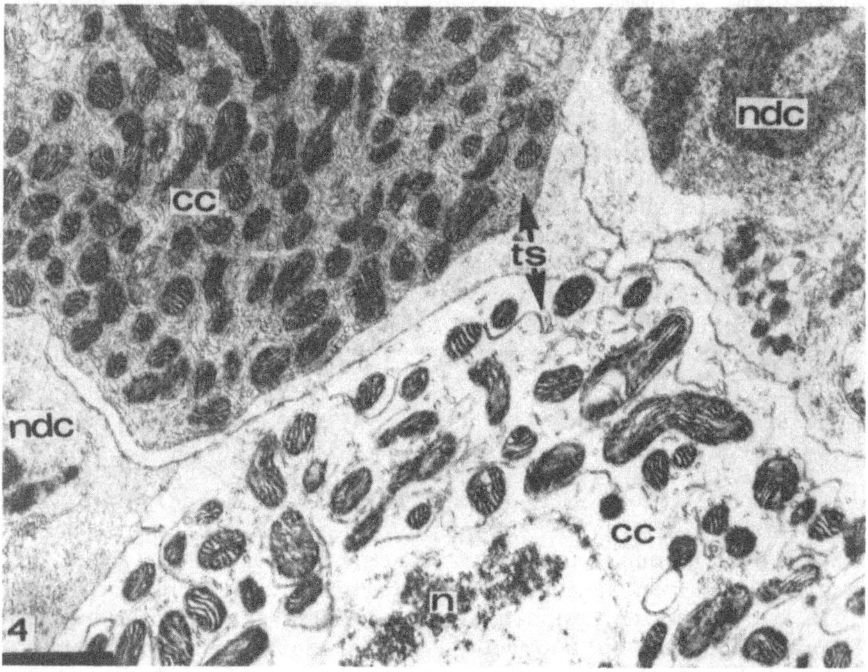

Figure 2. Gill cells from Mugil chelo adapted to 200% seawater treated with an amount of P. marmoratus secretion equivalent to 2 LD_{50} (4 μg/mL/gram of body weight), which paralyzed the fish within 20 min. The most damaged chloride cell is the nearest to the external membrane (lower section of plate). Note the damage of chloride and nondifferentiated cells. Scale 3 μm. (Reprinted by permission from Primor et al., 1980.)

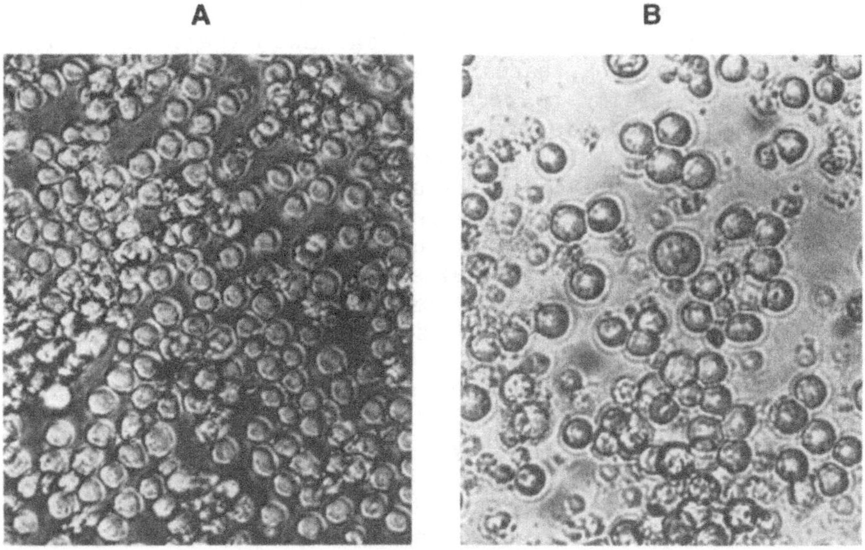

Figure 3. Tissue-cultured suspension of malignant mouse lymphoma cells treated with 100 μg/mL of pardaxin (A). At a very initial stage of envenomation, 15 min after application, indicating a more or less homogenous and continuous layer of cells. (x300) (B) Three hours later. Note empty spaces (indicate the disruption of cells); the presence of swollen hypertrophied cells (demonstrating vacuolization and granulations); and fragments of destroyed cells. (x300). (Taken from Lazarovici, unpublished)

when applied in large doses and/or prolonged applications, to more restricted and specific interferences with membranal functions that are mediated by proteins such as ionic channels, carrier molecules, and membrane-bound enzymes. This hypothesis--pardaxin's effect on membrane phospholipids--is supported by three separate and independent findings.

The first is actually an observation performed during the process of purification of pardaxin (Parness, 1975), which indicated that pardaxin acts as a surfactant and possesses the capacity to reduce water surface tension. This surface activity was expressed first of all by pardaxin's ability to cause a strong foaming of aqueous solutions and, secondly, by its ability to reduce the drop volume of aqueous solutions. The latter was shown when the collection of the eluent from a gel-filtration chromatography column was monitored by drop number. The collected volume in the successive tubes was progressively decreasing proportionally to the concentration of pardaxin eluted.

The second finding corresponds to most recent data concerning the effect of pardaxin on the permeability of artificial phospholipid-bilayer vesicles, the so-called liposomes (Barenholz et al., unpublished). This is a common preparation employed for the investigation of the chemical and physical properties of phospholipid bilayers. Liposomes are advantageous because they (a) are easily prepared; (b) possess a barrier that separates the internal medium from the external solution; and (c) represent phospholipid bilayers, which resemble their arrangement in biological membranes.

In our specific case, the integrity of the liposomes was monitored through the rate of release of a fluorescent compound (6-carboxy fluorescein, Blumenthal et al., 1977) by which they were preloaded. In the experiment described in Figure 4A, the liposomes were prepared from a synthetically modified lecithin (dipalmitoyl phosphatidylcholine, having its gel-to-liquid crystalline phase transition at 41°C). The concentration of pardaxin in the overall system was in the range of 0.2 - 0.3 μg/mL of medium. In this experiment the effect of pardaxin was compared with that of melittin--a low-molecular-weight, strongly basic and amphipatic protein derived from bee venom (Condrea, 1974; Knoppel et al., 1979), well-known to interact hydrophobically with phospholipid bilayers and biologic membranes and to possess cytolytic activity. Figure 4A shows that the ability of pardaxin to affect liposomes increases in time and is temperature-dependent. Melittin in the same concentration by weight (but about six times more concentrated on the basis of their molar ratio) was only slightly more active.

In the same series of experiments we also employed the so-called direct lytic factor of cobra venom (fraction D_5, derived from the venom of Naja mossambica, Primor and Zlotkin, 1978). In order to reach the same order of activities as shown by pardaxin and melittin, we submitted this substance 50 times (by weight) more concentrated. The data presented in Figure 4B indicate that the effect of pardaxin is basically independent in the chemical nature of the phospholipid and in its physical properties. In this experiment (Figure 4B) we employed two additional phospholipids strongly differing in their phase transition temperatures. Pardaxin was more effective than melittin in liposomes composed of bovine brain sphingomyelin. Therefore, pardaxin affects the integrity of phospholipid bilayers independently in the essence of the phospholipid; it is at least as active as melittin and is two-to-three orders of magnitude more active than a cobra venom direct lytic factor.

The third finding in support of the hypothesis on pardaxin's phospholipid interaction follows from its N-terminal amino acid sequence. Pardaxin is an acidic protein composed of 162 amino acids and molecular weight of 17,000 daltons. Its amino acid composition (Table 1) indicates approximately a 1:1 ratio between apolar-hydrophobic and polar amino acids, as shown by the vast majority of proteins. In the past we have also noticed a curious resemblance in the amino acid composition of pardaxin to those of a group of acidic and calcium-binding proteins, the so-called parvalbumins, present in the muscles of aquatic primitive vertebrates (Pechere et al., 1973; Primor et al., 1978). Table 2 presents the N-terminal sequence of the first 10 amino acids of pardaxin compared to those of melittin and several parvalbumins. In contrast to its amino acid composition (Table 1), the N-terminal region of the molecule demonstrates an extremely hydrophobic and positively charged (presence of lysine at step 8) nature, closely resembling melittin in this aspect (Table 2). Perhaps, since biologic membranes are in general negatively charged (in their protein as well as phospholipid components), the presence of such a positive charge may increase the proximity between the hydrophobic region of these molecules and the membranal phospholipids. In other words, the presence of an electrostatic affinity should amplify the hydrophobic interaction. This is a basic question in molecular biology that may be treated by the aid of substances such as pardaxin. The three different parvalbumins shown in Table 2 demonstrate a completely different sequence. The most obvious differences follow from their acetylated N-termini and the presence of several carboxylic polar groups.

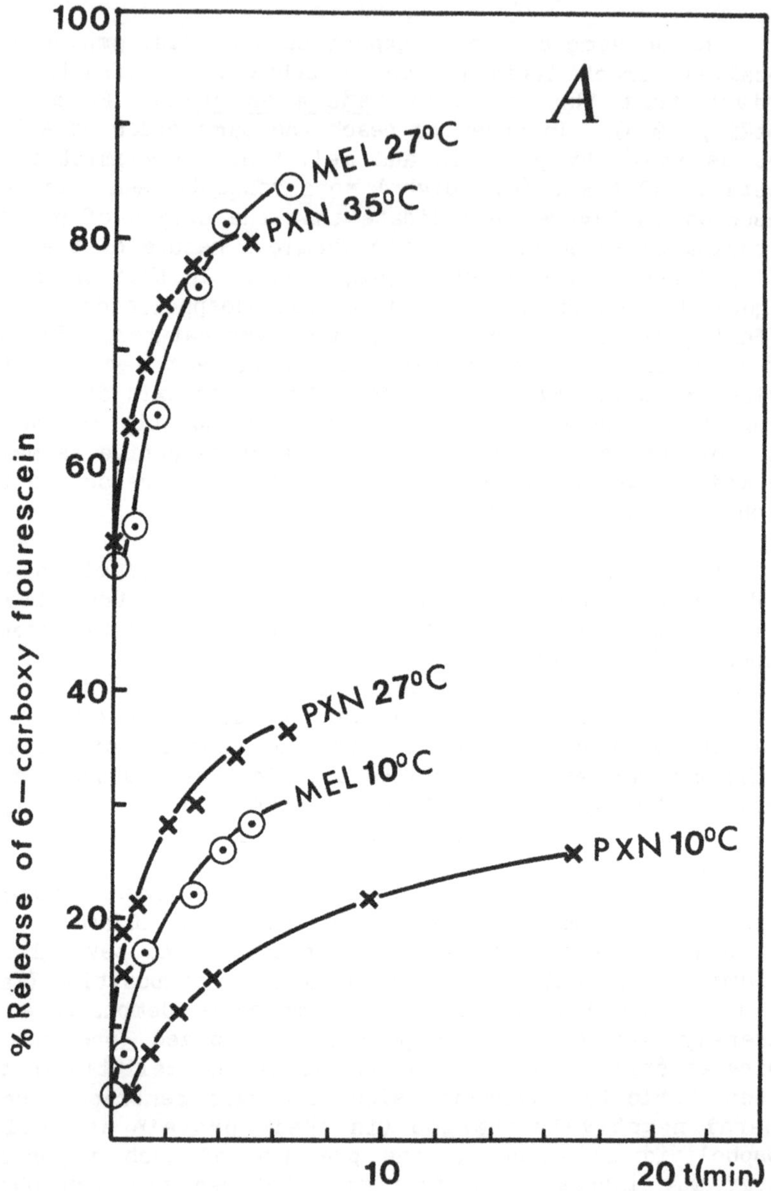

Figure 4. The effect of pardaxin (PXN) and melittin (MEL) on the release of liposome content. Unilamellar liposomes, prepared by sonication, were loaded by 6-carboxy fluorescein and purified by gel filtration. The fluorescence released was measured by a Perkin-Elmer spectrofluorimeter. The total release (100%) was determined by a treatment of liposomes with the detergent Triton x-100. Protein to lipid ratio was 1 to 20. The final concentrations of pardaxin and melittin were 0.2-0.3 g/mL of medium. (A.) Liposomes were composed of dipalmitoyl phosphatidyl choline (phase transition 41°C).

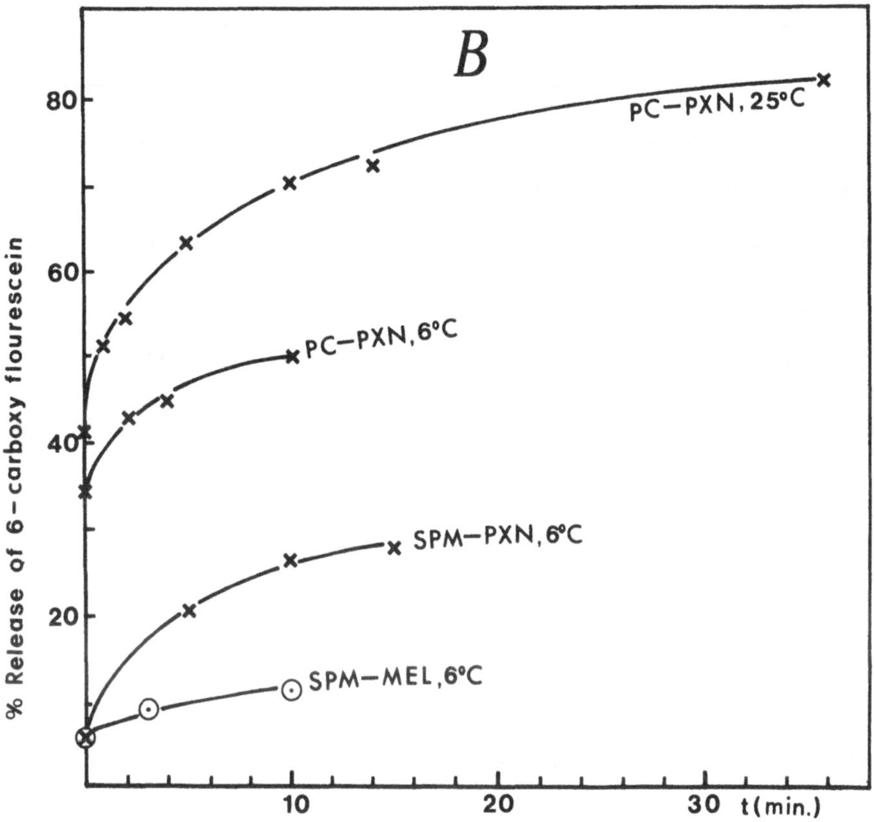

Figure 4. (B.) Liposomes were composed either of egg-phosphatidylcholine (PC-phase transition below 0°C) or bovine sphingomyelin (SPM-phase transition 28-40°C). (Taken from Barenholz, Cohen, and Zlotkin, unpublished)

Table 1. Amino acid composition of pardaxin (From Primor et al., 1978)

Amino Acid	No. of residues Experimental	Theoretical
Aspartic acid	8.99	9
Threonine	6.11	6
Serine	21.60	22
Glutamic acid	10.60	11
Proline	10.88	11
Glycine	14.89	15
Alanine	14.78	15
Half-cystine	6.77	8[c]
Valine	6.60	7
Mathionine	2.78	3
Isoleucine	11.22	11
Leucine	18.33	18
Tyrosine	1.95	2

Table 1, continued

Amino Acid	No. of residues Experimental	Theoretical
Phenylalanine	13.22	13
Lysine	10.99	11
Histidine	0[b]	0
Arginine	0[b]	0
Tryptophan	0[a]	0
TOTAL	159.71	162[d]

a. Based on spectrophotometrical estimations. The estimated mw of pardaxin--its micromolar extinction coefficient at 280 nm = 2.6, which indicates an absence of tryptophan and corresponds to 1.95 residues of tyrosine.

b. Not detectable, or amount too low to be quantified (less than 1n mole).

c. Estimated number based on previous experience, considering the lability and degradation.

d. The calculated molecular weight corresponds to 16,845.

Table 2. The N-terminal primary structure of pardaxin, melittin, and three parvalbumins.

Substance	Sequence Steps 1 2 3 4 5 6 7 8 9 10	Source
Pardaxin	H_2-Gly-Phe-Phe-Ala-Leu-Ile-Pro-Lys-Ile-Ile	Zlotkin et al., in preparation
Melittin	H_2-Gly-Ile-Gly-Ala-Val-Leu-Lys-Val-Leu-Thr	Haberman and Jentsch, 1967
Parvalb.-Pike	Ac-Ser-Phe-Ala-Gly-Leu-Leu-Lys-Asp-Ala-Asp	Gerday, 1976
Parvalb.-Carp	Ac-Ala-Phe-Ala-Gly-Val-Leu-Asp-Asp-Ala-Asp	Demaille et al., 1974
Parvalb.-Hake	Ac-Ala-Phe-Asp-Gly-Leu-Leu-Lys-Ala-Asp-Asp	Demaille et al., 1974

Concluding Remarks

Our hypothesis attributing the effects of pardaxin to an interaction with membranal lipid components is based on the following considerations: (1) The broad diversity and versatility is in the pharmacologic action of pardaxin; (2) it has surfactant capacities; (3) it can affect the integrity of liposomes; and (4) the strongly hydrophobic nature of its N-terminal sequence suggests an amphipatic character of its molecule. When dealing with the significance of these considerations from the point of view of this symposium, there are at least two aspects that deserve attention in the future:

1. It may be possible to employ pardaxin as a pharmacological tool for the study of protein-lipid interactions in biological membranes. For this purpose we intend to expand the studies concerning its structure/function relationships using a variety of preparations such as monolayers, bilayers, and simple models of biological membranes.

2. In the continuous and persistent search for shark repellents, it may be worthwhile to pay attention to surface-active substances, particularly their screening, characterization, and selection. Such substances are not expensive, simple, and available. In spite of the unsatisfactory data obtained by Office of Naval Research investigators with these substances in the early 1960s, the most recent series of experiments (performed by Zlotkin and Gruber in Miami, June 1981, unpublished), with various surfactants on lemon sharks, are promising and encouraging.

Acknowledgments

These studies were supported by grants from the U.S.-Israel Binational Science Foundation (BSF) Nos. 1730 and 2669, Jerusalem, Israel, Grant No. 14-75-C-0766, Office of Naval Research, U.S. Navy, and PHS NIH Grant HL 17576.

References Cited

Blumenthal, R., J.N. Weinstein, B.D. Sharron, and P. Kenkart. 1977. Liposomes-lymphocyte interaction: saturable sites for transfer and intracellular release of liposome contents. Proc. Natl. Acad. Sci. USA 74:5603.

Clark, E. 1974. The Red Sea's sharkproof fish. Natl. Geogr. 145(5):718-728.

Clark, E., and S. Chao. 1973. A toxic secretion from the

Red Sea flatfish Pardachirus marmoratus (Lacepede). Bull. Sea Fish Res. Stn. (Haifa) 60:53-56.

Condrea, E. 1974. Membrane-active polypeptides from snake venoms: cardiotoxins and haemocytotoxins. Experientia 30:121-129.

Demaille, J., E. Dutruge, J.P. Capony, and J.F. Pechere. 1974. Muscular parvalbumins. A family of homologous calcium-binding proteins. Their relation to the calcium-binding troponin component. Pages 643-675 in W. Drabikowski, H. Strzelecka-Golaszewska, and E.Carafoli, eds. Calcium-Binding Proteins. PWN-Polish Scientific Publishers, Warszawa.

Gerday, C. 1976. The primary structure of the parvalbumin II of pike (Esox lucius). Eur. J. Biochem. 70:305-318.

Haberman, E., and J. Jentsch. 1967. Sequenzanalyse des Melittins aus den tryptischen und peptischen Spaltstucke. Z. Physiol. Chem. 348:37-50.

Knoppel, E., D. Gisenberg, and W. Wickner. 1979. Interactions of melittin, a protein model, with detergents. Biochemistry 18:4177-4181.

Lavie, V. In preparation. Ph.D. Thesis, Hebrew University.

Parness, J. 1975. A study on the action and composition of the toxic secretion of the flatfish Pardachirus marmoratus (Soleidae). M.Sc. Thesis, Hebrew University.

Parness, J., and E. Zlotkin. 1976. Action of the toxic secretion of the flatfish Pardachirus marmoratus on the guinea-pig ileum. Toxicon 14:85-91.

Pechere, J.R., J.P. Capony, and J. Demaille. 1973. Evolutionary aspects of the structure of muscular parvalbumins. Syst. Zool. 22:533-544.

Primor, N., and E. Zlotkin. 1975. On the ichthyotoxic and hemolytic action of the skin secretion of the flatfish Pardachirus marmoratus (Soleidae). Toxicon 13:227-231.

_____. 1978. Oral toxicity of venoms and toxins to blowflies. Pages 1087-1095 in P. Rosenberg, ed. Toxins: Animal, Plant, and Microbial. Pergamon Press, Oxford.

Primor, N., J. Parness, and E. Zlotkin. 1978. Pardaxin: the toxic factor from the skin secretion of the flatfish Pardachirus marmoratus (Soleidae). Pages 539-547 in P. Rosenberg, ed. Toxins: Animal, Plant, and Microbial. Pergamon Press, Oxford.

Primor, N., I. Sabnay, V. Lavie, and E. Zlotkin. 1980. Toxicity to fish, effect on gill ATPase and gill ultra-structural changes induced by Pardachirus secretion and its derived toxin pardaxin. J. Exp. Zool. 211:33-43.

Spira, M.E., M. Klein, B. Hochner, Y. Yarom, and M. Castel. 1976. Ultrastructural changes accompanying the disturbances of neuromuscular transmission caused by Pardachirus toxin. Neuroscience 1:117-124.

Zlotkin, E., N. Primor, F. Couraud, and P. Lazarovici. 1979. Studies on the structure and mode of action of the ichthyotoxin pardaxin. Progress Report Abstracts. Office of Naval Research, Report ACR 230, pp.101-102, Arlington, Va.

Zlotkin, E., C. Kopeyan, and H. Rochat. In preparation. N-terminal primary sequence of pardaxin.

Naftali Primor, Jose A. Zadunaisky,
Anthony T. Tu

9. Pardaxin: Its Action, Toxicity, and Structure

In the 19th century ichthyologists noted that soles of the genus Pardachirus (Soleidae) possess a series of more than 200 pores along their dorsal and anal fins. Pressing this area would release a milky fluid from the fish (Klunzinger, 1871; Lacépède, 1802). Only recently Clark and Chao (1973) and Clark and George (1979) demonstrated that when suspended in water, this fluid is lethal to fishes. Consequently, Clark suggested that its biological function is to protect the flatfish from its predators. Indeed, a predatory shark responded to the fluid by avoidance, indicating its defensive function (Clark, 1974).

The toxicity of the secretion has been attributed to a single protein of a molecular weight of 14,000 daltons (Primor et al., 1978). Named pardaxin, this toxin is constructed of a monomeric chain, with a helical structure and four disulfide bridges (Primor and Tu, 1980).

In teleosts and elasmobranchs the gills are the most probable site of pardaxin's action (Primor et al., 1980a, b). Moreover, studies of several investigators indicate multiple pharmacological effects of the secretion, including effects on nerve-muscle tissue (Parness and Zlotkin, 1976; Spira et al., 1976) and the formation of channels in a lipid bilayer membrane (Korchak, 1979; Moran et al., 1977) and in the virion membrane (Pal et al., 1981a, b). We shall review here some recent findings of pardaxin-induced biological activities, emphasizing which are and which are not direct consequences of its capability to increase membrane leak.

In the teleost Fundulus heteroclitus a dose of 3 to 4 X LD_{50} administered into the medium caused an immediate response in the tested fish. They struggled and jumped sporadically, were unable to perform normal opercular movements, and the opercular rate decreased. Death occurred within 6-8 min.

Investigators using a chamber apparatus that separated the fish anterior from the rest of its body found that toxicity to F. heteroclitus occurred when flatfish secretion and pardaxin were administered to the head part of the chamber. The secretion's disruption of fish gill membrane (Primor et al., 1980a) further indicated that they were the target of toxicity. These observations indicated that the surface of the head region of the fish is the target for pardaxin-induced responses. Since ion-osmoregulatory properties of the opercular epithelium closely resemble those of the teleost gills (Degnan et al., 1977), it was used to study the mechanism of pardaxin's action in fish.

The Effect of Pardaxin on Electrical Properties of the Short-Circuited Killifish (F. heteroclitus) Operculum Epithelium

Administered at 120 μg/mL into the mucosal (seawater) side of the opercular epithelium, pardaxin elicited a dual effect on the short-circuited current. At first, the short-circuited current was slightly stimulated and then, within 20 min, mostly inhibited. At 40 μg/mL the short-circuited current recovered within 80 min, but at higher concentrations it was irreversibly inhibited. Following pardaxin action, the epithelium was capable of responding to isoprenaline stimulation (5×10^{-7} M) in the serosal side. This finding suggests that pardaxin does not destroy the epithelium's response to β-adrenergic activation of Cl^- secretion. In order to identify the ion(s) that might move across the epithelium causing stimulation of the short-circuited current, experiments were run in which a specific ion was removed and a substitution made.

In Cl^--free Ringer, pardaxin (120 μg/mL) administered into the mucosal side elicited a large short-circuited current stimulation. This effect was not observed when pardaxin was added to the serosal side. In the operculum epithelium, the rate of Cl^- secretion was dependent upon the HCO_3^- concentration (Degnan et al., 1977). In HCO_3^--free Ringer, pardaxin (120 μg/mL) administered to the mucosal side produced short-circuited current stimulation similar to that observed in Cl^--free Ringer.

In Na^+-free Ringer, pardaxin (120 μg/mL in the mucosal side) failed to stimulate the short-circuited current. Removal of K^+ resulted in short-circuited current decrease (from 180 $\mu A/cm^2$ to 30 $\mu A/cm^2$); however, pardaxin was able to stimulate it.

Pretreatment with ouabain (10^{-5}M in the serosal side), the classical inhibitor of Na^+, K^+-ATPase, abolished the short-circuited current stimulation induced by pardaxin (120 μg/mL in the mucosal side) in both the Cl^--rich and the Cl^--free Ringer. The simplest interpretation of this experiment is that ouabain abolishes the ion-gradient between the cells and the bathing medium and prevents a net sodium movement across the epithelium. These experiments indicate that Na^+ is directly involved in pardaxin-induced short-circuited current stimulation, but that Cl^-, HCO_3^-, and K^+ are not.

Within 20 min following the administration of pardaxin (120 μg/mL in the mucosal side), the net flux of Cl^- was inhibited by 48.8%. The opercular epithelium displayed large unidirectional sodium fluxes but no significant net flux (Degnan and Zadunaisky, 1980). Pardaxin inhibited the sodium efflux, which resulted in a significant (P = 0.005-0.01) net sodium influx of 2.2 μequiv. x cm^{-2} x hr^{-1}.

The unidirectional inulin fluxes were four to five orders of magnitude lower than the sodium fluxes. Within 20 min, the inulin unidirectional fluxes reached a maximum of 10-15 times higher than the control (from 0.16 x 10^{-3} μequiv. x cm^{-2} x hr^{-1} to 2.2 x 10^{-3} μequiv. x cm^{-2} x hr^{-1}) and then decayed. They were still 3.3 times higher than the control after 50 min.

Toxicity of P. marmoratus Secretion and Pardaxin to Fish and Other Animals

Although the toxicity of P. marmoratus secretion was discovered by its administration into the medium (Clark and Chao, 1973), a comparison of this toxicity to that produced by injection into the body of the animal is significant. The secretion and pardaxin show low toxicity when injected into the bloodstream of A. dispar, as compared to their toxicity when administered into the medium (Primor et al., 1980a). A comparative study of toxicity for an elasmobranch had similar results. Using a group of 40 small dogfish (Squalus acanthias) weighing 40-46 g each, the lethal doses (LD_{50}) of the secretion and pardaxin were 8.0 and 5.1 μg/mL/g body weight, respectively, as determined one hour after administration into 250 mL of seawater. An injection into a dorsal artery, however, required much higher doses of 90.0 and 54.0 μg/g body weight for the secretion and pardaxin, respectively, as determined after one hour (Primor et al., 1980b). This one order of magnitude difference in toxicity demonstrates a bias in pardaxin's action on the mucosal (seawater) surface of the gills. Table 1 shows the quantitative determinations of the toxicities of the secretion and pardaxin to

Table 1. Toxicity to fishes by *Pardachirus* secretion (PMC) and pardaxin (PX) administered into the media.

Fish Species Weight(g)	Lethal Dose (LD_{50}) (μg/mL of medium /g body weight) PMC	PX	Medium Salinity	Source
Gambusia affinis (0.7-0.12)	10.5	nt[a]	freshwater	Primor and Zlotkin, 1975
Lebistes reticulatis (0.5-0.1)	4.2	2.9	freshwater	Primor, unpublished
Aphanius dispar (0.5-1.5)	17.3	nt	deionized water	
	108.0	nt	30% seawater	Primor et al., 1980
	12.5	9.0	150% seawater	
	3.3	nt	250% seawater	
Mugil chelo (3-5)	2.0	nt	200% seawater	Primor et al., 1980
(150-160)	0.30	nt	seawater	Primor, unpublished
Fundulus heteroclitus (2-3)	42.0	30.0	seawater	in preparation
	1000	500.0	fish Ringers	
		32.0	seawater	
Squalus acanthias (40-46)	8.0	5.1	seawater	Primor et al., 1980b
Rhizoprinodon terraenovae (150)	∼0.13	nt	seawater	Primor et al., 1982
Pardachirus marmoratus (150-180)	>2.6	nt	seawater	Primor, unpublished
Bothus sp. (150-200)	∼0.25	nt	seawater	Primor, unpublished

a. Not tested.

fish, when administered into the medium.

The secretion's toxicity to A. dispar and F. heteroclitus depended on the salinity of the medium so that a high salinity environment produced a greater sensitivity to it. This suggests an ability of pardaxin to disrupt the osmoregulatory processes occurring in the gills.

We also examined the possibility that the secretion was toxic to species other than fishes. In tadpoles of the green toad Bufo viridis the toxicity of the secretion and pardaxin administered to the pond water was 15.7 and 11.8 μg/mL/0.1 g body weight, respectively. A dose of 3-4 x LD_{50} administered into the water caused a rapid response: an intense struggling similar to that observed in teleosts. Death occurred within 3 to 5 min. Pardaxin injected into the rat vein almost immediately decreased the arterial blood pressure. The LD_{50} of the secretion and pardaxin by injection into the rat femoral vein was 8.5 and 4.6 mg/kg, respectively (Primor and Lazarovici, 1981). When the toxicity of the secretion was tested for the crustaceans Macrobrachium rosenbergii and Artemia salina, they survived a concentration of 3 mg/mL (Primor, unpublished). This unexpected finding, given that crustaceans (like fish) are dependent upon gill-osmoregulating structures, requires additional study.

When Pardachirus fish weighing approximately 150 g were introduced into seawater containing 400 μg/mL of the secretion, they survived indefinitely, even though this challenging dosage was 10 times greater than the lethal dose for another species of flatfish (Bothus sp.) of similar weight. Thus, Pardachirus shows a definite resistance to its own secretion. However, after administration of the substance, its gill structure showed a serious lysis in gill membrane. This finding eliminates the possibility that Pardachirus gills are resistant to the cytolytic action of its secretion (Primor, unpublished). The mechanism of the self-immunization of Pardachirus to its own secretion deserves further study.

The Effect of the Secretion and Pardaxin on Transmitter Release and the Lipid Bilayers

The action of the secretion on transmitter release and muscle contraction was studied in the frog's cutaneous pectoris nerve-muscle preparation and the ileum smooth-muscle preparation. In the neuromuscular preparation, its primary action was to increase the number of neurotransmitter quanta and the frequency of mEPP. However, observed muscle tremors in the presence of tubocurarine and morphological disruptions of the muscle fibers also indicated a direct effect on the contractile mechanisms (Spira et al., 1976).

In a guinea-pig ileum preparation, the secretion (10 μg/mL) induced a contractile response of the ileum and inhibited the stimulation induced by 5-hydroxytryptamine (5-HT). A muscle contraction was interpreted as the secretion's action on nerve terminals, resulting in a release of cholinergic transmitter (Parness and Zlotkin, 1976).

Fractionation of the secretion resulted in four major proteinic fractions, which were individually tested on the ileum contraction and inhibition to 5-HT stimulation. Pardaxin (2 μg/mL) caused a potent muscle contraction and inhibited the response to 5-HT stimulation (Primor, in preparation). Since the receptors stimulated by 5-HT are located on the nerve cell membrane (Primor, 1980), the finding suggests an action of pardaxin on nerve membrane resulting in inhibition of transmitter release.

The planar lipid membranes served as a model system to study the nature of the interaction between the P. marmoratus secretion and lipid bilayers. At a concentration of 0.1 μg/mL there was a significant increase in the lipid conductance (Moran et al., 1977). The secretion induced several distinct changes in the electrical characteristics of the lipid bilayer membrane: (a) The current-voltage curves showed a significant rectification, indicating that the increase in the membrane conductance was voltage-dependent; this was observed when the toxin was administered to the positive side of the Lucide chamber. (b) Threshold voltage was dependent upon the ion-strength of the solution; in concentrated KCl solution the increase in membrane conductance was highly elevated. (c) At a concentration of 0.001 μg/mL the membrane conductance increased in a discrete uniform step, indicating that the mechanism of secretion interaction with lipid membranes involves the formation of transmembranal channels. (d) The membrane activity is proportional to the sixth power of the toxin concentration, indicating the secretion's ability to anchor in the lipid bilayers an aggregate form of six molecules. And (e) there is no apparent selectivity in conductance to different ions (Korchak, 1979).

This study was performed with the secretion; however, pardaxin is probably responsible for the secretion's activity increasing the lipid membrane conductance. Indeed, pardaxin makes lipid bilayer vesicle membranes permeable, as evidenced by leakage of entrapped 6-carboxyfluorescein (Zlotkin and Barenholz, chapter 8 of this volume). These studies support the notion that the cytolytic action of the toxin works directly on the lipid bilayer component of the cell membrane. Pardaxin's ability to disrupt lipid bilayer in biological systems has also been demonstrated in the vesicular stomatitis virus. At 50-100 μg/mg virus it caused activation of the transcriptase activity of the virion, presumably by an increase in the envelope permeability. However, at a concentration of 200-300 μg/mg virus, a dissociation of the virion lipid envelope from the nucleocapsid occurs (Pal et al., 1981b).

Properties of Crude P. marmoratus Secretion and Pardaxin

Secretion subjected to gel chromatography on Sephadex G-10 resulted in high- and low-molecular-weight fractions. The low-molecular-weight fraction contained a factor capable of neutralizing the hemolytic as well as toxic activity of the secretion (Primor et al., 1975). The chemical nature of this antitoxin compound is still unclear. The major component of the secretion, pardaxin, represents about 40% of its total proteins; the secretion contains several other proteins, one of which exhibits hemolytic activity (Primor and Zlotkin, 1975). When submitted to acrylamide gel electrophoresis, the secretion shows several protein bands. The toxicity was located in the most acidic protein (Primor and Zlotkin, 1976). No enzymatic activity has yet been identified.

The secretion possesses neither esterase nor phospholipase A_1 or A_2 activities (Primorand Lazarovici, 1981). When submitted to chromatography on a Sephadex G-25 column, a reduction of the eluent drop volume in the high-molecular-weight fraction occurred. Volume measuring of the elution tubes of column Sephadex G-100 indicated that the volume reduction occurs within an elution profile of several proteins and is not attributable to pardaxin alone (Primor, in preparation). This finding indicates that pardaxin reduces the liquid surface tension. Direct measurements of the surface tension by Cenco-du-Nouy surface-tension balance shows that the secretion (at 1 mg/mL) reduced the water surface-tension by 60% (Primor, unpublished). The effect of both the secretion and pardaxin on the surface tension and its possible significance requires additional study.

Two steps of fractionation by column gel filtration chromatography resulted in the product named pardaxin (Primor et al., 1978). Its homogeneity was established by disc gel electrophoresis, and the molecular weight of 13,100 g/mL was estimated by SDS method. SDS gel electrophoresis of a mercaptoethanol-treated pardaxin yielded only one protein band (Primor, in preparation). This finding, as well as the lack of Raman vibration in the range of 100-200 cm^{-1}, strongly suggest that pardaxin is composed of a single polypeptide chain. Its secondary structure was determined by two methods: laser Raman and circular dichroism (Primor and Tu, 1980).

The Raman spectrum of pardaxin has a prominent amide I and amide III at 1658 cm^{-1} and 1280 cm^{-1}, respectively, which suggests the presence of α-helical structure. Examination of the far-ultraviolet spectrum showed a positive band at 196 nm and a negative at 222 nm. The overall spectrum qualitatively resembled the model spectra of α-helical polypeptides. The percentages of α-helical conformation as analyzed from Raman spectra indicated that pardaxin's structure is composed of 38.6% α-helix, 22.8% β-sheet, and 38.6% random coil. The best fit of pardaxin spectra based on circular dichroism ellipticity was 23% α-helix, 21% β-structure, and 56% random coil.

Examination of Raman spectra of pardaxin in 9% NaCl indicated that there is no substantial change in the peptide backbone conformation whether in water or in salt. This finding indicates that the increased pardaxin toxicity in high salinity is probably not an effect of its conformation resulting from pardaxin's activation.

When Raman spectroscopy was used to examine the conformation of disulfide bonds, the disulfide stretching vibration of pardaxin appeared at 510-516 cm^{-1}, indicating that its disulfide bridges are of gauche-gauche-gauche configuration. The lack of a sulfhydryl vibrational band in the 2500-2700 cm^{-1} region further suggested that all eight half-cystines which were identified (Primor et al., 1978) participated in disulfide bonds. It is still unknown which chemical characteristics are responsible for pardaxin-induced biological activity.

Summary and Discussion

In the operculum preparation pardaxin produced a net Na^+ influx of about 60.0 $\mu A/cm^2$; the net $^{36}Cl^-$ flux was inhibited within 48.8%. These findings indicate that pardaxin did not cause a substantial leak to Na^+ or to Cl^-. The inulin rate of permeability is very low (4 to 5 orders of magnitude lower than Na^+), so that even a small increase in its leak can be readily shown. Pardaxin-induced leak to inulin is about three orders of magnitude lower than the value of the net Na^+ influx it created. Therefore pardaxin action increasing the epithelial leak is only a small fraction of its activity; the most markable effect is the production of a net Na^+ influx.

The studies reviewed indicate multiple biological effects of pardaxin: the production of channels in lipid bilayer membranes; the effects on transmitter release from the presynaptic nerve terminals (Spira et al., 1976); skeletal and smooth muscle contraction (Spira et al., 1976; Parness and Zlotkin, 1976); inhibition of ATPase (Primor et al., 1980a); and disruption of cell membranes (Primor and Lazarovici, 1981; Pal et al., 1981a, b). There are at least two possible interpretations of such effects: They are all sequelae of permeability changes produced by the toxin-induced channels, or they are consequences of toxin interaction with specific component(s) of the membrane. As to the first possibility, the toxin-induced channels may depolarize the presynaptic nerve terminals leading to release of transmitter. Moreover, Ca^{2+} could enter through these channels in the muscle membrane causing muscle contraction. Leakage of electrolytes from cells might lead to the cytolytic effects. Indeed, lysis of cells which lack Na^+, K^+ -ATPase (Primor and Lazarovici, 1981) indicates that the effect on the cell membrane is not mediated by this specific membrane component. Finally, the results with respect to lipid bilayers might lead to prediction of a nonspecific toxic effect on biological membranes. With regard to the second possibility, in the operculum epithelium pardaxin induces a net Na^+ influx by inhibiting the Na^+ efflux, suggesting some specific interaction with a cell membrane.

Information gained from the experiments with opercular epithelium may resolve this dilemma. In spite of a general leak to inulin, pardaxin inhibited the Na^+ efflux and consequently produced a net Na^+ influx. In a high salinity medium the Na^+ influx can be even larger. Pardaxin's enhanced activity in a high salinity medium (Primor et al., 1980a) supports the notion that the production of a net Na^+ influx is the mechanism of pardaxin-induced behavioral responses and toxicity in fish. However, the question of how it acts on the mucosal side resulting in a net influx of sodium is still unresolved.

The search for surfactants to serve as repellents against sharks was preceded by the study of holothurin (Sobotka, 1965). This principal toxin of a Bahamian holothurian is a highly amphipatic molecule with a hydrophilic and a hydrophobic pole. In a dilution of 1:600,000, holothurin killed a 20 kg shark; smaller fish were killed at even greater dilutions.

In spite of an entirely different chemical identity, pardaxin and holothurin both display the characteristics of surfactants. Furthermore, some commercial surfactants also repel sharks (Gruber, 1982). However, it remains to be further studied whether the surfactant property of pardaxin is responsible for its various effects.

Acknowledgments

This research was supported by NIH Research Grant PHS/EY 01340, NIH Research Grant PHS/GM 25002, NIH Research Grant PHS/EY 03151, and Research Grant R01 GM 19172, the Department of Defense, Office of Naval Research, Research Grants No. N00014-80-C-0757 and N00014-81-C-0667.

References Cited

Clark, E. 1974. The Red Sea's sharkproof fish. Natl. Geogr. 145(5):718-728.

Clark, E., and S. Chao. 1973. A toxic secretion from the Red Sea flatfish Pardachirus marmoratus (Lacepede). Bull. Sea Fish Res. Stn. (Haifa) 60:53-56.

Clark, E., and A. George. 1979. Toxic soles, Pardachirus marmoratus from Red Sea and P. pavoninus from Japan, with notes on other species. Environ. Biol. Fishes 4(2):103-123.

Degnan, K.J., and J.A. Zadunaisky. 1979. Open-circuit sodium and chloride fluxes across isolated opercular epithelia from the teleost Fundulus heteroclitus. J. Physiol. 294:483-495.

_____. 1980. Ionic contributions to the potential and current across the opercular epithelium. Am. J. Physiol. 7: R231-R239.

Degnan, K.J., K.J. Karnaky, and J.A. Zadunaisky. 1977. Active chloride transport in the in vitro opercular skin of a teleost (Fundulus heteroclitus), a gill-like epithelium rich in chloride cells. J. Physiol. 271:155-191.

Gruber, S.H. 1982. Shark repellents: perspectives for the future. Oceanus 24(4):72-76.

Klunzinger, C.B. 1871. Synopsis der Fishe des Roten Meeres. Verh. Zool. Bot. Ges. Wien. 21:441-688.

Korchak, Z. 1979. The electrochemical characteristics of lipid bilayer modified with Pardachirus toxin. MSc. Thesis. Technion-Israel Institute of Technology, Haifa, Israel.

Lacépède, C. 1802. Histoire naturalle des poissons. 4 Chez Plasson, Imprimeur-Libraire, Paris.

Moran, A., N. Primor, and E. Zlotkin. 1977. The effect of Pardachirus venom on the conductance of lipid bilayer. Annual Meeting of the Israeli Pharmacological-Physiological Society, Nes-Ziona, Israel. (Abstract)

Pal, R., Y. Barenholz, and R.R. Wagner. 1981a. Pardaxin, a hydrophobic toxin of the Red Sea flatfish disassembles the intact membrane of vasicular stomatitis virus. J. Biol. Chem. 256:10209-10212.

_____. 1981b. Transcription of vesicular stomatitis virus activated by pardaxin, a fish toxin that permeablized the virion membrane. J. Virol. 39:641-645.

Parness, J., and E. Zlotkin. 1976. Action of the toxic skin secretion of the flatfish Pardachirus marmoratus on the guinea-pig ileum. Toxicon 14:85-91.

Primor, N. 1980. Inhibition of 5-hydroxytryptamine contraction of guinea-pig smooth muscle treated with black widow spider venom. Eur. J. Pharmacol. 68:497-500.

Primor, N., and P. Lazarovici. 1981. Pardachirus marmoratus (Red Sea flatfish) secretion and its isolated toxic fraction pardaxin: the relationship between hymolysis and ATPase inhibition. Toxicon 19:573-578.

Primor, N., and A.T. Tu. 1980. Conformation of pardaxin, the toxin of the flatfish Pardachirus marmoratus. Biochim. Biophys. Acta 626:229-306.

Primor, N., and E. Zlotkin. 1975. On the ichthyotoxic and hemolytic action of the skin secretion of the flatfish Pardachirus marmoratus (Soleidae). Toxicon 13:227-231.

_____. 1976. The toxic and antitoxic factors derived from the skin secretion of the flatfish Pardachirus marmoratus (Soleidae). Pages 287-294 in A. Ohsaka, K. Hayashi, and Y. Sawai, eds. Animal, Plant, and Microbial Toxins, Vol.1. Plenum Press, New York.

Primor, N., T. Zerachia, and E. Zlotkin. 1975. An antitoxic factor derived from the skin secretion of the flatfish Pardachirus marmoratus (Soleidae). Toxicon 13:183-187.

Primor, N., J. Parness, and E. Zlotkin. 1978. Pardaxin: the toxic factor from the skin secretion of the flatfish Pardachirus marmoratus (Soleidae). Pages 539-547 in P. Rosenberg, ed. Toxins: Animal, Plant, and Microbial. Pergamon Press, Oxford.

Primor, N., I. Sabnay, V. Lavie, and E. Zlotkin. 1980a. Toxicity to fish, effect on gill ATPase and gill ultrastructural changes induced by Pardachirus secretion and its derived toxin pardaxin. J. Exp. Zool. 211: 33-43.

Primor, N., J.A. Zadunaisky, V.H. Murdaugh and J.N. Forrest. 1980b. A study of pardaxin toxicity in the dogfish. Mt. Desert Is. Biol. Lab. (MDIBL) 20:120-122.

Sobotka, H. 1965. Comparative biochemistry of marine animals. BioScience 15:583-584.

Spira, M.E., M. Klein, B. Hochner, Y. Yarom, and M. Castel. 1976. Ultrastructural changes accompanying the disturbances of neuromuscular transmission caused by Pardachirus toxin. Neuroscience 1:117-124.

10. Toxicity in Shallow Marine Waters: Potential for Developing Shark Repellents

Yamanouchi in 1942 and Nigrelli in 1952 independently discovered that certain holothurians contained a toxin; both named it holothurin. Yamanouchi concluded that holothurian poison has little ecological significance (see historical reviews in Baslow, 1969; and Bakus, 1973). Sobotka (1965) believed that holothurin evidently protects sea cucumbers against predators, but he did not conduct experiments under natural conditions to support this conclusion. However, he did find one holothurin (from *Actinopyga agassizi* Selenka, 1867) that killed a 22-kilogram lemon shark in a dilution of 1:600,000 within 50 minutes. Scientists at that time did not know that numerous other species of tropical marine animals (e.g., sponges, soft corals) possess toxins, nor did they pursue further the use of holothurin as a shark repellent. I independently discovered holothurin through field studies in 1967 and later demonstrated the ecological significance of this and other marine toxins.

Although my first underwater observations of coral reefs occurred north of Puerto Limon, Atlantic Costa Rica, in 1962, it was during the Fanning Island Expedition of June-August 1963 that the effects of fish feeding activities on the benthic biota became apparent. Fanning Island is a coral atoll, one of the Line Islands, in the central equatorial Pacific (4°N, 159°W). I had planned to conduct experiments on the effects of currents on sponge growth and form, but I could not locate exposed, large branching sponges for my research (one species was later found in the atoll lagoon). Underwater observations showed that large numbers of fish (e.g., parrotfish, wrasses, damselfish, surgeonfish) were feeding on the dead coral substratum and its biota, and numerous *in situ* experiments verified the effects of fish on the benthic biota (Figure 1). I concluded that grazing and predacious fish probably have played an important role by acting as agents of natural selection, leading to the

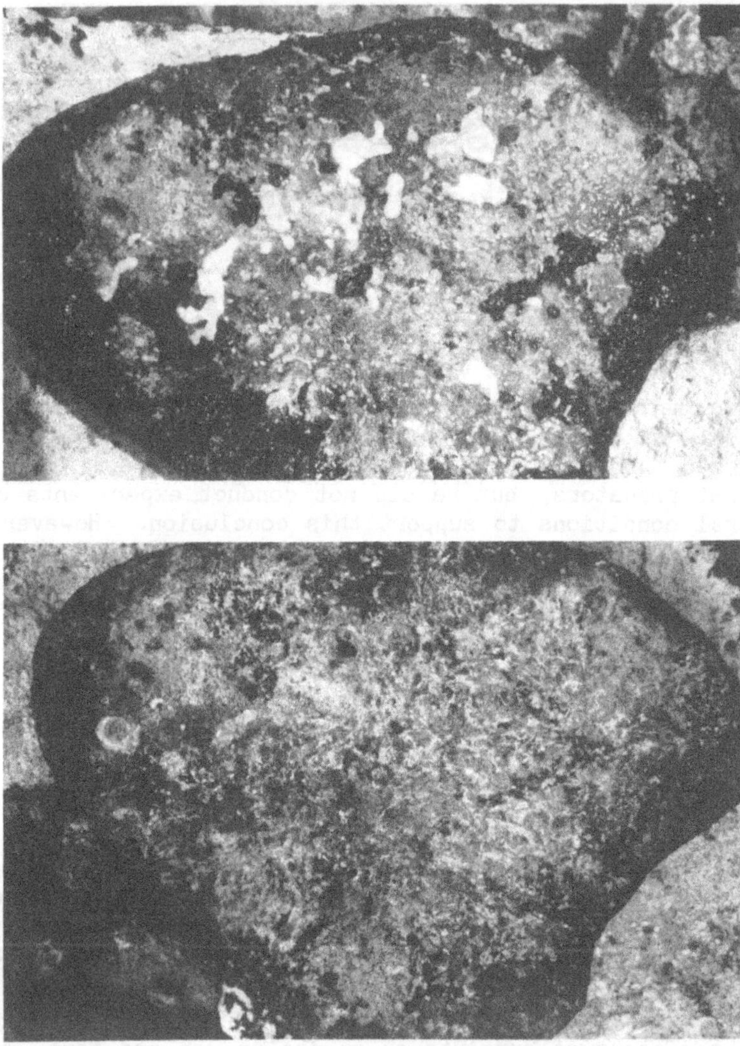

Figure 1. Top: View of the underside of a coral slab collected from 1 m of water at Fanning Island, Line Islands, central Pacific. Note the dark sponges (*Aplysilla*, *Halisarca*, Dendroceratida), light didemnid ascidians (*Didemnid*), and peripheral dark alga (*Pocockiella*). Bottom: The same coral slab after placing it in 0.6 m of water with the underside exposed to rasping fishes for only 1 hr. Note the numerous toothmarks and the remnants of a bivalve mollusc (scar at center left).

evolution of protective mechanisms and habits among certain tropical shallow-water invertebrates (Bakus, 1964). I included toxicity among the various types of protective mechanisms considered, but carried out no experiments at that time.

In 1965, I attempted to repeat Fanning Island-type field experiments at Enewetak Island, Marshall Islands, but did not find similar circumstances (e.g., coral slabs that could be readily transplanted from regions of high-intensity fish predation and grazing to relatively low-intensity fish-feeding sites) to conduct comparable studies. However, the general absence of exposed benthic biota (excepting hard corals, soft corals and sea cucumbers) was again apparent, which led me to studies on the defense mechanisms and ecology of selected species of holothurians (sea cucumbers) at Enewetak Island in 1967. Of particular interest was a small (4 cm by 1 cm), brown holothurian (<u>Holothuria</u> <u>difficilis</u> Semper, 1868) that was living exposed and abundant (mean = 21 individuals/900 cm^2 on slab rock at a depth of less than or equal to 1 m), especially at the edge of a quarry (Bakus, 1968). It was perplexing that such a bite-size species could survive with so many fish and sharks living and feeding in the same area.

Experiments showed that all the major body parts (i.e, body wall, viscera) were highly toxic to reef fish and that Cuvierian organs (tubules) from these holothurians were not only ejectable on provocation, but that they too were toxic. Surprisingly, the ejection of tubules (generally considered a defense mechanism, discussed in Bakus, 1973) occurred only during the day, when the large majority of predacious and grazing coral reef fish are active (they "sleep" at night). This suggested that such a temporal behavioral pattern may have evolved in response to high-intensity predation by fish (including sharks) during the daytime. Other species of exposed holothurians were also toxic to fish. Moreover, the only sizeable species of exposed sponge in the Enewetak Island quarry was also toxic to fish. This suggested that chemical defense mechanisms are important in the survival of numerous reef organisms (Figure 2).

In a subsequent article, I discussed in detail the literature relevant to patterns of species richness and feeding habits of fish with latitude (Bakus, 1969, Table 1). I described protective and defensive mechanisms by which exposed coral reef invertebrates could receive protection from fish, including large size, skeletal and tube protection, cryptic habits (living within and between the branches of corals), evasive movement, nocturnal habits, cryptic

Figure 2. Rubbing of the body surface of the holothurian Holothuria atra Jaeger, 1833 results in the release of a burgundy-colored cloud, consisting of the pigment echinochrome and the toxin holothurin. Holothurin rapidly kills small tidepool fish and some invertebrates. Interestingly, the sea cucumber may carry a black (cryptic) polynoid polychaete as a commensal on its body surface. The polychaete dies when holothurin is released.

Table 1. Characteristics of marine fishes at different latitudes (from various authors, in Bakus, 1969)

Characteristic	Remarks
Species Richness	
Western Arctic America to Eastern Arctic Siberia, nearshore and deep waters	50 species
Southern California, 0-30 m	180 species
Hawaiian Islands, nearshore and deep waters	600 species estimated
Seychelle Islands, nearshore and deep waters	900 species estimated
South Australia, nearshore and deep waters	341 species
Eastern Antarctic, 0-100 m	17 species
Feeding Habits	
Southern California kelp beds	45 species[a]
Marshall Islands coral reefs	212 species[b]
Standing Crops	
Southern California kelp beds	(0.035 kg fish/m^2)
Hawaii coral reefs	(0.185 kg fish/m^2)

a. Herbivores, 7%; omnivores, 44%; carnivores, 49%.

b. Herbivores, 25%; omnivores, 8%; carnivores, 67%.

coloration and form, and noxious, poisonous, venomous, and aposematic defense mechanisms. In the Caribbean I discovered further examples of toxic sponges and holothurians (Bakus, 1972). There was a high incidence of cryptofauna--hidden animals, so characteristic of corals (see McCloskey, 1970)--within coral reefs, which I attributed to predation and grazing pressure by fish. This may have led to shifts in gene frequency in which the phototaxes (swimming and settling behavior) of numerous marine larvae had presumably changed from positive (giving rise to exposed adults) to negative (giving rise to the cryptofauna) as a result of natural selection. Further studies on toxic sponges and holothurians from as far north as the San Juan Islands (Washington) to Cocos Island (tropical eastern Pacific), combined with information from the literature, showed that toxicity in these taxa to a variety of marine and freshwater test fish (i.e., goldfish) is inversely related to latitude and may reach 100% for holothurians in high-diversity coral reefs (Bakus, 1974, 1975; Bakus and Green, 1974). Moreover, feeding experiments in the field with species of fish co-occurring with the toxic invertebrate species confirmed results from the laboratory.

During a Scripps Alpha Helix cruise in March 1978 between Belize and Cozumel, Mexico, I collected numerous marine organisms. Bioassays on 54 species (70 specimens) of representative Caribbean sponges revealed 31 species (57%) toxic to fish. There was excellent correspondence between my own simple 20-minute bioassay and standard 24-hr LC_{50} experiments. The fish exhibited a diversity of behavioral reactions to sponge toxins, which I interpreted as responses to the complexity of the chemistry of marine sponges (Bakus and Thun, 1979).

Lizard Island, north Great Barrier Reef, Australia, was the site of further studies on toxic benthic marine organisms (Bakus, 1981). Because this is one of the most species-rich areas in the world, I expected a relatively high degree of toxicity. Of all exposed common coral reef invertebrates tested (42 species)--comprising sponges, soft corals, gorgonians, seastars, crinoids, holothurians, and ascidians--73% were toxic to fish. Most of the remaining exposed species had structural defense mechanisms against predation by fish (e.g., hard skeletons, mucus). Results from feeding experiments conducted in the field at Lizard Island, combined with recent laboratory experiments described in the literature, suggested that marine fish may learn to avoid toxic organisms by trial-and-error feeding.

We now know many soft corals and crinoids from the Great Barrier Reef that are toxic to fish (Coll et al.,

1982; M.D. Sutherland, personal communication). Scleractinian corals are generally nontoxic probably because they are protected by nematocysts and hard skeletons (J.C. Coll, personal communication). Some marine sponges and soft corals release toxins regularly into seawater in measurable amounts (Janice Thompson; J.C. Coll, personal communications). Further information on marine toxins described in the recent literature is in Bakus (1981) and in papers by chemists, especially those at the University of Oklahoma, Scripps Institution of Oceanography, University of California at Santa Cruz, and the University of Hawaii. Table 2 presents a summary of information on the toxicity of shallow-water benthic marine organisms.

I tested crude extracts of toxic Caribbean sponges on sharks at our marine station. First, Gerardo Green and his laboratory assistant, Patricia Gomez, prepared a crude extract from pieces of red sponge (*Haliclona rubens*) known to contain the fish toxin halitoxin. The sponge specimens had been collected from the Caribbean coast of Mexico and frozen; the mass of sponge pieces extracted approximated 10 cm in diameter, and the crude extract had a volume of about 150 mL. The crude extract was frozen for 6 months before testing.

Greg Pittinger captured two swell sharks (*Cephaloscyllium ventriosum*) off Santa Catalina Island, California, in February 1981 and maintained them in a running seawater aquarium at the University of Southern California Marine Science Center. On 27 March 1981 the sharks (~0.75 m in length) were placed in separate aquaria (~60 liters seawater each) and became active after handling, later settling motionless on the bottom, respiring once every 3 seconds. Approximately 50 mL of the sponge crude extract was diluted 1:6 in fresh seawater. A turkey baster (50mL) was filled with seawater then emptied underwater near the snout of the first shark (time 0) without a response. The turkey baster was then filled with seawater-diluted crude extract and emptied in the same manner. The shark began active body thrashing in 7 seconds and continued this behavior for 185 seconds, after which it settled on the bottom of the aquarium. It began to elicit convulsive respirations, its mouth agape and forcefully swallowing seawater every 15 to 20 seconds, slowing to every 30 seconds after 9 minutes. Respiration returned to normal after 15 minutes.

The turkey baster was rinsed in seawater then filled with fresh seawater and emptied underwater near the snout of the second shark without a reaction. The turkey baster was then filled with seawater-diluted crude extract and emptied as before (time 0). The shark began active body thrashing in

Table 2. Toxicity of shallow-water marine organisms at various latitudes[a]

Locality	Latitude	Taxonomic Group	No. of Species Tested	No. of Species Toxic	% of Species	Source
Canada	48°N to 78°N	freshwater and marine fish	?	12 (of 770)[b]	1.6	McAllister, 1968
San Juan Islands, Washington	48°N	sponges	34	3	9	Bakus and Green, 1974; Green, 1977
		holothurians	12	3	25	Bakus, 1974; Bakus and Green, 1974
Nontropics	less than 20°N & S	marine organisms	937	262[b]	28	Various authors, in Bakus, 1969
Onagawa, Japan	38°N	holothurians	5	4	80	Yamanouchi, 1955
Seto, Japan	35°N	holothurians	9	7	78	Yamanouchi, 1955
Santa Catalina Island, California	33°N	sponges	44	9	21	Bakus and Green, 1974; Green, 1977
		holothurians	2	1	50	Bakus, 1974; Bakus and Green, 1974

Table 2. (Cont.)

Locality	Latitude	Taxonomic Group	No. of Species Tested	No. of Species Toxic	% of Species	Source
Guaymas, Mexico	28°N	holothurians	6	5	83	Bakus, 1974; Bakus and Green, 1974
Zihuatanejo Bay, Guerrero, Mexico	17°N	sponges	11	7	64	Bakus and Green 1974; Green, 1977
La Blanquilla Reef, Veracruz, Mexico	19°N	sponges	36	27	75	Bakus and Green 1974; Green, 1977
Tropics	Approx. 20°N to 20°S	marine organisms	937	675[b]	72	Various authors, in Bakus, 1969
Cozumel, Mexico, to Belize, Belize	20°N to 17°N	sponges	54	31	57	Bakus and Thun, 1979
Eniwetok, Marshall Islands	12°N	holothurians	4	4	100	Bakus, 1968

Table 2. (Cont.)

Locality	Latitude	Taxonomic Group	No. of Species Tested	No. of Species Toxic	% of Species	Source
Palau Islands, Pacific Ocean	7°N	holothurians	11	11	100	Yamanouchi, 1955
Cocos Island, eastern Pacific	6°N	holothurians	7	6	86	Bakus, 1974
Lizard Island, north Great Barrier Reef, Australia	14°S	sponges, soft corals, gorgonians, asteroids, criniods, holothurians, ascidians	42	25	60[c]	Bakus, 1981

a. Organisms toxic to fish unless otherwise indicated.

b. Organism toxic to humans.

c. 73% of exposed benthic species.

8 seconds. When the remainder of the diluted crude extract was added to the aquarium, convulsive respirations began after 4.5 minutes and the shark continued thrashing for 15 minutes, then settled on the bottom of the aquarium. Respiration ceased after 17 minutes, and the shark appeared to be nearly dead after 20 minutes. When it was prodded and its body bent, there was no response. Only its eyes would slowly close when the eyelid was touched with the end of the turkey baster, the body partly swollen with swallowed seawater. Respiration was still absent after 30 minutes. The aquarium was drained, while fresh seawater was added. The shark began moving and elicited convulsive respirations after 33 minutes. Both sharks recovered fully from the experiments (Greg Pittinger, personal communication).

On 3 April 1981, Greg Pittinger, Jeff Landesman, Paul Bakus, and I traveled 5 kilometers northwest of Ship Rock, off Santa Catalina Island, California, where we attracted blue sharks (Prionace glauca) after 44 minutes by suspending cut mackerel in a steel mesh cage. Two pieces of mackerel (4 x 4 x 2 cm) were injected and perfused with sponge crude extract containing halitoxin. We fed mackerel to the sharks by dropping a large piece over the side of the boat and observed their feeding behavior through a concave-convex glass bubble mounted in the floor of the boat. Then we fed them two pieces of halitoxin mackerel at different times. They ate both the thawed mackerel and halitoxin mackerel in the same manner, simply swallowing the pieces of fish whole. When the sharks approached the side of the boat, we made several attempts to squirt halitoxin crude extract diluted in seawater into their snouts, but elicited no response.

Discussion

Evidence from approximately 1000 laboratory and field experiments and underwater observations in many regions of the world indicates that the intensity of fish and shark predation and rasping on the shallow-water hard benthos and the incidence of toxic benthic marine invertebrates vary inversely with latitude. These two phenomena probably represent cause and effect. An examination of the literature on fish- and shark-feeding habits indicates that sharks rarely eat toxic benthic invertebrates, but exceptions do occur. For example, a toxic holothurian found near the reef edge at Heron Island, Australia (Stichopus chloronotus Brandt, 1835), is occasionally eaten by sharks (Bakus, 1968). Of course, many large predators are opportunists that will consume most anything when they are hungry (Bakus, 1969); even pelagic species (e.g., tiger sharks) may eat benthic organisms on occasion. Although rotenone (used to collect fish) is deadly to fishes and sharks in sufficient concentration, hungry jacks and whitetip sharks will commonly swim through a cloud of rotenone to eat a thrashing fish affected by the poison. Although no sharks of danger to man have heretofore been subjected to sponge toxins (as far as is known), they would probably be repelled by such toxins in sufficient concentration; after all, rotenone is a terrestrial plant toxin, whereas these marine toxins probably have evolved in the presence of sharks and other vertebrate predators. However, the sharks more dangerous to man are relatively large (usually close to 2 meters long or longer) and the amount of toxin needed to deter a hungry individual might be considerable.

Among the organisms highly toxic to fish are several sponges and holothurians. The most toxic Caribbean sponge known (Hymeniacidon ?amphilecta) caused cessation of opercular movement in test fish in only 3 minutes. A 24-hr LC_{50} value for this sponge was 1.4 μg/mL (test fish = goldfish); a 96-hr LC_{50} value was 0.8 μg/mL (Thun, 1980). This may be compared with a 96-hr LC_{50} value of 16 μg/mL (test fish = mosquitofish) for the toxin pardaxin from the Red Sea flatfish Pardachirus marmoratus (Moses sole) (Primor and Zlotkin, 1976). These values and those from several other Porifera suggest that certain marine sponges may be among the most toxic marine organisms known; thus, their toxins may be very effective shark repellents.

Recent tests conducted on swell sharks show that invertebrate toxins that affect teleost fish tend to affect the sharks in a similar manner. Halitoxin may produce an immediate effect similar to that of pardaxin from the Moses sole, that is, by altering gill membrane structure and interfering with oxygen transport. It may also have other long-term effects such as lysis of red blood cells. Although laboratory experiments produced marked behavioral changes in swell sharks, halitoxin had no effect on blue sharks because it was diluted very rapidly in the somewhat choppy seas and the sharks quickly swallowed whole pieces of fish. Even squirting it near the snout of the shark was unsuccessful because of the rapid dilution of the toxin. If invertebrate toxins are to be very effective as shark repellents, some mechanism must be developed to keep them from dispersing too rapidly in open seas.

The problem with a variety of shark repellents (e.g., chemical repellents, electric shock, black bags, bubble barriers, sonic barriers) is that they are either unreliable, expensive, ineffective against many species of sharks, or not easily used by individuals. Because of dilution in the open sea, even the most toxic chemicals have proved only marginally effective. For example, Shark Chaser used by the U.S. Navy, consisting of 80% black nigrosine dye and 20% copper acetate, is effective against several species of sharks but not Caribbean nurse sharks or Pacific blacktips (Anonymous, 1971; Baldridge, 1969; Gilbert and Gilbert, 1973; Klimley and Myrberg, 1979).

The technique of finding suitable shark repellents should be to collect known benthic invertebrates that are the most highly toxic to fish and test crude extracts from them with a few species of sharks. Collections of additional toxic species from selected tropical regions would be advisable since (a) each species of dangerous shark would probably respond somewhat differently to a given toxin, and (b) even more effective shark repellents might be discovered. The crude extracts producing the greatest deterrent to sharks could then be treated to produce a series of chemical fractions, which could be bioassayed to identify specific toxic components. The toxic fractions could be tested with selected species of dangerous sharks; combined fractions could then be tested for their effects on shark behavior, and the most effective ones could be combined with Shark Chaser. Available chemical carriers for these toxic fractions could be used to provide the best characteristics of suspension and dispersion of toxins to deter shark attacks under given oceanographic conditions.

Finally, the most promising compounds could be chemically synthesized to produce the toxins in volume or in concentration. The plethora of toxic marine organisms in tropical coral reef ecosystems would make the probability of discovering effective shark repellents in the near future very high. However, the most difficult problems to overcome include the prevention of rapid chemical dilution of repellents in the open sea and the prevention of shark attacks in progress.

Acknowledgments

Greatly appreciated is the collection of sharks, vessel, and field assistance provided by Greg Pittinger, Jeff Landesman, and Donald Nelson. I thank Douglass W. Birch and Richard Brusca, Allan Hancock Foundation, University of Southern California; Bernard Zahuranec, Office of Naval Research; and a perceptive critic for reviewing the manuscript.

References Cited

Anonymous. 1971. Electric shocks effectively repel sharks. Austr. Fish. 30(11):14-15.

Bakus, G.J. 1964. The effects of fish-grazing on invertebrate evolution in shallow tropical waters. Allan Hancock Found. Occ. Pap. (27):1-29.

_____. 1968. Defense mechanisms and ecology of some tropical holothurians. Mar. Biol. 2(1):23-32.

_____. 1969. Energetics and feeding in shallow marine water. Intl. Rev. Gen. Exp. Zool. 4:275-369.

_____, ed. 1972. Marine studies on the north coast of Jamaica. Atoll Res. Bull. 152:1-6.

_____. 1973. The biology and ecology of tropical holothurians. Pages 325-367 in O.A. Jones and R. Endean, eds. Biology and Geology of Coral Reefs, Vol. 2: Biology 1. Academic Press, New York.

_____. 1974. Toxicity in holothurians: a geographic pattern. Biotropica 6:229-236.

_____. 1975. Marine zonation and ecology of Cocos Island, off Central America. Atoll Res. Bull. 179:1-9.

_____. 1981. Chemical defense mechanisms and fish feeding behavior on the Great Barrier Reef, Australia. Science 211:497-499.

Bakus, G.J., and G. Green. 1974. Toxicity in sponges and holothurians: a geographic pattern. Science 185:951-953.

Bakus, G.J., and M. Thun. 1979. Bioassays on the toxicity of Caribbean sponges. Colloq. Intl. Centre Natl. Recherche Sci. (291):417-422.

Baldridge, H.D., Jr. 1969. Analytic indication of the impracticability of incapacitating an attacking shark by exposure to waterborne drugs. Mil. Med. 134(12): 1450-1453.

Baslow, M.H. 1969. Marine Pharmacology: A Study of Toxins and Other Biologically Active Substances of Marine Origin. Williams and Wilkins, Baltimore.

Coll, J.C., S. La Barre, P.W. Sammarco, W.T. Williams, and G.J. Bakus. 1982. Chemical defences in soft corals (Coelenterata: Octocorallia) of the Great Barrier Reef: a study of comparative toxicities. Mar. Ecol. Prog. Ser. 8:271-278.

Gilbert, P.W., and C.K. Gilbert. 1973. Sharks and shark deterrents. Underwater J. 5(2):69-79.

Green, G. 1977. Ecology of toxicity in marine sponges. Mar. Biol. 40:207-215.

Klimley, A.P., and A.A. Myrberg, Jr. 1979. Acoustic stimuli underlying withdrawal from a sound source by adult lemon sharks, Negaprion brevirostris (Poey). Bull. Mar. Sci. 29(4):447-458.

McAllister, D.E. 1968. Poisonous and venomous fishes of Canada. Natl. Mus. Canada Nat. Hist. Papers (42):1-11.

McCloskey, L.R. 1970. The dynamics of the community associated with a marine scleractinian coral. Intl. Rev. Ges. Hydrobiol. 55(1):13-81.

Primor, N., and E.Zlotkin. 1976. The toxic and antitoxic factors derived from the skin secretion of the flatfish Pardachirus marmoratus (Soleidae). Pages 287-294 in A. Ohsaka, K. Hayashi, and Y. Sawai, eds. Animal, Plant, and Microbial Toxins, Vol. 1. Plenum Press, New York.

Sobotka, H. 1965. Comparative biochemistry of marine animals. BioScience 15:583-584.

Thun, M.A. 1980. Bioassays on the toxicity of Caribbean sponges. Master's Thesis. Department of Biological Sciences, University of Sourthern California.

Yamonouchi, T. 1955. On the poisonous substance contained in holothurians. Pubs. Seto Mar. Biol. Lab. 4:183-203.

Index

For Product Safety Concerns and Information please contact our EU representative GPSR@taylorandfrancis.com Taylor & Francis Verlag GmbH, Kaufingerstraße 24, 80331 München, Germany